Out-of-School-Time STEM Programs for Females

A volume in
Out-of-School-Time STEM Programs for Females:
Implications for Research and Practice
Lynda R. Wiest, *Series Editor*

Out-of-School-Time STEM Programs for Females

Implications for Research and Practice

Volume I: Longer-Term Programs

edited by

Lynda R. Wiest
University of Nevada, Reno

Jafeth E. Sanchez
University of Nevada, Reno

Heather Glynn Crawford-Ferre
University of Nevada, Reno

INFORMATION AGE PUBLISHING, INC.
Charlotte, NC • www.infoagepub.com

Library of Congress Cataloging-in-Publication Data

A CIP record for this book is available from the Library of Congress
http://www.loc.gov

ISBN: 978-1-68123-843-2 (Paperback)
978-1-68123-844-9 (Hardcover)
978-1-68123-845-6 (ebook)

Cover photo by Lynda R. Wiest

CONTENTS

INTRODUCTION

Why OST STEM Programs for Females?

**Lynda R. Wiest, Jafeth E. Sanchez,
and Heather Glynn Crawford-Ferre**

The science, technology, engineering, and mathematics (STEM) disciplines play a pivotal role in societal progress and economic prosperity (Riegle-Crumb, King, Grodsky, & Muller, 2012). Their contributions to societal well-being, coupled with the poor performance of U.S. students internationally, has led to calls for building a more STEM-proficient workforce (Davis & Hardin, 2013; Dillivan & Dillivan, 2014). STEM fields possess high status and offer career promise (e.g., stable, growing, high-paying occupations), providing important employment opportunities that can influence people's life quality (Riegle-Crumb et al., 2012; Shapiro, Grossman, Carter, Martin, Deyton, & Hammer, 2015). It is thus disconcerting that females show less interest and confidence in STEM (Ross, Scott, & Bruce, 2012; Shapiro et al., 2015) and are underrepresented in STEM pursuits, including career preparation through persistence in careers, as well as recreational involvement, such as participation in clubs and contests (Wiest, 2011).

Out-of-school-time (OST) programs, which are on the rise, are increasingly suggested as one strategy for supporting and encouraging females in STEM (Afterschool Alliance, 2013; McCombs et al., 2012). These

Out-of-School-Time STEM Programs for Females, pages ix–xii
Copyright © 2017 by Information Age Publishing

programs offer less formal but structured learning environments ranging from a few hours to a week or longer during summer, weekend, and after-school time (Mohr-Schroeder et al., 2014; Wilkerson & Haden, 2014). OST STEM programs are typically designed for participants to explore STEM content and careers, apply STEM to real-world settings and develop awareness of the utilitarian value of STEM, and inspire interest in STEM (Afterschool Alliance, 2013; Mohr-Schroeder et al., 2014; Wilkerson & Haden, 2014). Accordingly, participants in OST STEM programs have shown improved achievement, interest, and confidence in STEM and greater awareness of STEM role models and careers (McCombs et al., 2012; Mohr-Schroeder et al., 2014; Shapiro et al., 2015). Ways to address the content vary and can include, for example, enhancing, augmenting, or integrating content (White, 2013), but use of experiential learning is common across programs (Bevan & Michalchik, 2013; Dillivan & Dillivan, 2014; Mohr-Schroeder et al., 2014). Programs of greater intensity and duration have been found to be more effective for learning and retention (Mosatche, Matloff-Nieves, Kekelis, & Lawner, 2013).

The chapters in this volume detail OST STEM programs for females that run one week or more in length. Table I.1 provides an overview of the featured programs. (Note that an unsuccessful effort was made to locate contributing authors for programs outside the United States.)

The seven programs that appear in this book range from one week to one year in length. Four are commuter (day) programs, and three are residential (overnight) programs. Most are held on college campuses. Program participants range from upper elementary through high school students, with 25–60 girls attending each program. All are female, but the type of girl targeted differs (e.g., underserved girls, girls with a particular background or interest in STEM, or a random mix). The staff, who tend to be educators and students, are either all female or mostly female. The cost to participants for the sample programs in this book is either free (five programs) or several hundred dollars with tiered costs based on family financial status (two programs). Various types of internal and external support are integral to program sustainability.

In this book, the chapter authors describe their programs, the effectiveness of those programs, and practical implications of data they have collected on their programs. Terminology note: Despite older conceptions of sex as biological and gender as psychosocial, the term *gender* has increasingly replaced *sex* in discussions of females and males in the social sciences (e.g., Haig, 2004; Hubbard, 1996) and thus is the term of choice in this book.

TABLE I.1 Overview of Programs Featured in This Book

Name	Year Began[a]	Location	Main STEM Focus	Website
Las Chicas de Matemáticas	1999	Colorado	Math	http://www.unco.edu/nhs/mathsci/mathcamp/
Matherscize	1999	Ohio	Math	http://u.osu.edu/erchick.1/matherscize-3/
Techbridge	2000	California, Washington (state), Washington, DC	Science, technology, engineering	www.techbridgegirls.org
Eureka!-STEM	2012	Nebraska	Science, technology, engineering, math	N/A (embedded in UNO STEM Outreach page at http://www.unomaha.edu/college-of-education/office-of-stem-education/community-engagement/index.php)
Northern Nevada Girls Math & Technology Program	1998	Nevada	Math, technology	http://www.unr.edu/girls-math-camp
GOALS for Girls	2008	New York	Science, technology, engineering	https://www.intrepidmuseum.org/GOALSforGirls
All Girls/All Math	1997	Nebraska	Math	http://www.math.unl.edu/programs/agam

[a] "Year Began" means the first year the program was conducted, although it might have been developed the previous year.

REFERENCES

Afterschool Alliance. (2013). *Defining youth outcomes for STEM learning in afterschool.* Retrieved from http://www.afterschoolalliance.org/STEM_Outcomes_2013.pdf

Bevan, B., & Michalchik, V. (2013, Spring). Where it gets interesting: Competing models of STEM learning after school. *Afterschool Matters,* 1–8.

Davis, K. B., & Hardin, S. E. (2013). Making STEM fun: How to organize a STEM camp. *TEACHING Exceptional Children, 45*(4), 60–67.

Dillivan, K. D., & Dillivan, M. N. (2014). Student interest in STEM disciplines: Results from a summer day camp. *Journal of Extension, 52*(1), 1–11.

Haig, D. (2004). The inexorable rise of gender and the decline of sex: Social change in academic titles, 1945–2001. *Archives of Sexual Behavior, 33*(2), 87–96.

Hubbard, R. (1996). Gender and genitals: Constructs of sex and gender. *Social Text, 14*(1/2), 157–165.

McCombs, J. S., Augustine, C., Schwartz, H., Bodilly, S., McInnis, B., Lichter, D., & Cross, A. B. (2012). Making summer count: How summer programs can boost children's learning. *The Education Digest, 77*(6), 47–52.

Mohr-Schroeder, M. J., Jackson, C., Miller, M., Walcott, B., Little, D. L., Speler, L., . . . Schroeder, D. C. (2014). Developing middle school students' interests in STEM via summer learning experiences: See Blue STEM Camp. *School Science and Mathematics, 114*(6), 291–301.

Mosatche, H. S., Matloff-Nieves, S., Kekelis, L., & Lawner, E. K. (2013, Spring). Effective STEM programs for adolescent girls: Three approaches and many lessons learned. *Afterschool Matters,* 17–25. Retrieved from files.eric.ed.gov/fulltext/EJ1003839.pdf

Riegle-Crumb, C., King, B., Grodsky, E., & Muller, C. (2012). The more things change, the more they stay the same? Prior achievement fails to explain gender inequality in entry into STEM college majors over time. *American Educational Research Journal, 49*(6), 1048–1073.

Ross, J. A., Scott, G., & Bruce, C. D. (2012). The gender confidence gap in fractions knowledge: Gender differences in student belief-achievement relationships. *School Science and Mathematics, 112*(5), 278–288.

Shapiro, M., Grossman, D., Carter, S., Martin, K., Deyton, P., & Hammer, D. (2015). Middle school girls and the "leaky pipeline" to leadership. *Middle School Journal, 46*(5), 3–13.

White, D. W. (2013). Urban STEM Education: A unique summer experience. *Technology and Engineering Teacher, 72*(5), 8–13.

Wiest, L. R. (2011). Females in mathematics: Still on the road to parity. In B. Atweh, M. Graven, W. Secada, & P. Valero (Eds.), *Mapping equity and quality in mathematics education* (pp. 325–339). New York, NY: Springer.

Wilkerson, S. B., & Haden, C. M. (2014). Effective practices for evaluating STEM out-of-school time programs. *Afterschool Matters,* 19, 10–19.

CHAPTER 1

A DOOR TO STEM POSSIBILITIES

Las Chicas de Matemáticas

Hortensia Soto-Johnson

ABSTRACT

The lack of women in STEM-related careers is a growing concern, especially in light of the projected need for STEM professionals. Las Chicas de Matemáticas, a free residential camp, is designed to address this concern by promoting STEM-related careers to high school girls. Due to its prominent role in all STEM fields, college-level mathematics is the focus of the camp. This chapter contains a description of the structure of the camp, as well as the manner in which the program influences young women's confidence in their ability to do mathematics, informs them about STEM-related careers, and piques their interest in learning advanced mathematics. Of particular interest is how this challenging content, taught through inquiry-based learning, promotes creativity and is instrumental in furthering the participants' interest in mathematics. This has instructional implications for other in- and out-of-school activities related to STEM.

Out-of-School-Time STEM Programs for Females, pages 1–28
Copyright © 2017 by Information Age Publishing

> *Each of us has that right, that possibility, to invent ourselves daily.*
> *If a person does not invent herself, she will be invented.*
> *So, to be bodacious enough to invent ourselves is wise.*
>
> —Maya Angelou

For decades, there has been concern with the lack of women pursuing and completing science, technology, engineering, and mathematics (STEM)-related careers. It is especially disconcerting that women's disinterest in STEM fields begins at an early age. The literature indicates that although girls tend to perform as well as boys in elementary school, they start to doubt their mathematics and science abilities in middle school at about age twelve (Commission on the Advancement of Women and Minorities in Science, Engineering, and Technology Development, 2000; Eccles, 1989). This lack of confidence persists through high school (Eccles, 1989) and might be one factor that influences the low representation of women in college STEM courses (National Science Board, 2006). It is especially alarming that women who do pursue and complete STEM-related careers are not well represented in academia and in industry. For example, although women earn 41% of PhD's in STEM fields, they make up only 28% of tenure-track faculty in those fields (Beede, Julian, Langdon, McKittrick, Khan, & Doms, 2011). Further, recent reports show that men make up two-thirds of the staff for Google, Facebook, and other major technology companies (Guynn, 2014).

These dire statistics can be attributed to many factors. For instance, during their adolescent years girls tend to suffer from low self-esteem, which contributes to poor academic performance and suppressed ambitions (Jobe, 2002/2003). Unfortunately, both parents and teachers unintentionally cultivate this low self-esteem in young women. For example, parents tend to attribute their son's success in mathematics and science to natural talent and their daughter's success to effort and hard work (Aschbacher, Li, & Roth, 2010; Eccles, 1989). Moreover, parents tend to view mathematics as more important for the future of their sons than their daughters. While young women tend to articulate a desire to pursue careers that help and nurture people, most parents are not well-informed about STEM-related careers that focus on the well-being of others, such as humanitarian engineering. Thus, parents often fail to encourage their daughters to pursue more mathematics and science coursework.

Some teachers also contribute to gender bias in STEM. In particular, they fail to provide young women with role models, who can be instrumental in helping young women envision themselves succeeding in STEM-related careers (Seymour, 2006; Seymour & Hewitt, 1997). Given these realities, it is no wonder that a 2011 report by the Executive Office of the President of the United States promotes support systems that attract women into and retain them inthe STEM workforce. Such support systems may include mentoring by successful women in STEM-related careers, opportunities to engage in STEM-related activities, and encouragement to pursue STEM-related careers. These support systems must begin as early as elementary school so that young girls pursue advanced STEM-related courses and extracurricular activities during their middle and high school years. These support systems are also imperative during and after college, especially because women do not secure a high percentage of STEM jobs in academia or in industry.

The primary purpose of this chapter is to describe the program Las Chicas de Matemáticas, which is designed to promote STEM-related careers to high school girls. Given the prominent role that mathematics plays in all STEM fields, the program focuses on mathematics. A secondary aim of this chapter is to report how this program influences young women's confidence in their ability to do mathematics, informs them of STEM-related careers, and piques their interest in learning advanced mathematics. The delivery of this program is informed by research on dispositions and motivation toward learning mathematics, learning environments that inspire women to pursue STEM-related careers, and characteristics of successful outreach programs.

RELATED LITERATURE

In his summary of research on student dispositions toward mathematics, Beyers (2011) found varied interpretations of the notion of disposition, which he categorized into three constructs. From mathematics education literature, Beyers identified "the existence of dispositional cognitive, affective, and conative mental functions that contribute to a student's mathematical disposition" (p. 71). Cognitive dispositional mental functions with respect to mathematics include perceiving, recognizing, conceiving, judging, reasoning, and similar mental actions. Affective dispositions focus on one's identity with mathematics, that is, beliefs about oneself as a mathematics learner, beliefs about the value and usefulness of mathematics, and attitudes toward mathematics. The last disposition construct, conative, is associated with amount of effort, persistence, and diligence that one dedicates to learning mathematics. Teachers, role models, and peers, as well as significant family members, can influence students' disposition in all three of these constructs (Aschbacher et al., 2010; Beyers, 2011; Eccles, 1989; Jobe, 2002/2003) and, consequently, students' mathematics learning.

Soto-Johnson, Craviotto, and Parker's (2011) work suggests that the interplay among motivation, environmental factors, and disposition toward mathematics can influence whether mathematically high-performing young women pursue advanced mathematics. In their study, environmental factors such as relationships with parents, teachers, and peers, as well as participation in extracurricular mathematics activities, such as MathCounts, motivated the high school girls to do well in their mathematics courses. Such environmental factors and participation in extracurricular activities also influenced the high school girls' positive, confident disposition toward mathematics. Consequently, this enthusiastic disposition toward mathematics motivated the girls to do well and to create career goals that require a strong mathematics background. Finally, the high school girls' motivation to do well, combined with their mathematics-related career goals, promoted a positive, confident disposition toward mathematics. Thus, there is a strong interplay between disposition toward mathematics and motivation to do well in mathematics. In an effort to motivate and improve young women's disposition toward mathematics, Soto-Johnson et al. (2011) recommend that teachers "create learning environments where students can collaborate with one another and learn mathematics conceptually" (p. 138). This message is echoed in other literature (e.g., Eccles, 1989; Jobe, 2002/2003; Laursen, Hassi, Kogan, & Weston, 2014).

According to Eccles (1989), learning environments that are competitive, promote social comparisons, incorporate drill and practice, and minimize student–teacher interactions do not motivate middle and high school females to study mathematics and science. Eccles notes that learning environments that do benefit females use hands-on activities, non-sexist materials, cooperative formats that ensure full participation by all students, and career counseling. She says a noteworthy aspect of such learning environments is that they "facilitate the motivation and performance of minority students and low achieving males as well" (p. 53). Jobe (2002/2003) also found that reform teaching methods that emphasize verbal and visual problem-solving methods combined with teamwork benefit both girls and boys. Besides learning STEM content, both genders learn important interpersonal skills, such as cooperation, negotiation, conflict resolution, and listening. Similar results have been found with students enrolled in college-level mathematics courses (Laursen et al., 2014).

Laursen et al. (2014) investigated the impact of inquiry-based learning (IBL) in college mathematics courses. Teachers who taught using IBL methods dedicated over 60% of the class time to facilitating mathematics discussions through student-centered activities where the students presented their work, engaged in small-group work and discussions, and integrated technology. Teachers who taught using non-IBL methods spent 87% of the class time talking to the students. The research team found that students enrolled in IBL courses tended to show greater student leadership and ask more questions. Laursen et al. further noted:

Students in IBL math-track courses reported greater learning gains than their non-IBL peers on every measure: cognitive gains in understanding and thinking; affective gains in confidence, persistence, and positive attitude about mathematics; and collaborative gains in working with others, seeking help, and appreciating different perspectives. (p. 409)

In controlling for gender, the findings suggest that women enrolled in IBL courses had the same cognitive and affective gains as their male counterparts. The women's subsequent grades were also as good as their male peers, but their collaborative gains were higher than the males enrolled in the IBL courses. Women enrolled in non-IBL courses did not fare as well. They "reported substantially lower cognitive and affective gains than did their male classmates" (p. 411). Laursen et al. suggest that this particular result most likely reflects the women's weaker perception of their ability rather than a true difference in performance.

In sum, some literature suggests that IBL classroom environments favorably influence women's success with STEM curricula from kindergarten to college-level courses. Such teaching practices have produced similar positive results when integrated with outreach efforts in science (Laursen, Liston, Thiry, & Graf, 2007), technology (Matarić, Koenig, & Feil-Seifer, 2007), engineering (Weinberg, Pettibone, Thomas, Stephen, & Stein, 2007), and mathematics (Chacon & Soto-Johnson, 2003). For example, Laursen et al. (2007) argue that K–12 students become more interested in science and develop new perceptions about the field when they engage in hands-on activities with authentic scientists who visit the classroom. Through such visits, teachers also learn current scientific content and how to teach it creatively. Chacon and Soto-Johnson (2003) showed similar results for mathematics. The high school girls who participated in their summer program learned about the role of mathematics in STEM-related careers from women in STEM fields. These interactions, along with other camp activities, favorably influenced their perceptions of mathematics. Matarić et al. (2007) emphasized the importance of developing creative, accessible, and affordable STEM educational materials. Accordingly, they worked with K–12 schools to develop hands-on robotics courses, along with teacher training materials. Given that most teachers are not trained to teach robotics classes, the teacher materials facilitated implementation of the activities. Weinberg et al. (2007) found that girls who participated in a robotics program and who had good mentors "experienced an increase in self-concept and increased expectations of success in science and math due to the program" (p. 3). These researchers conclude that programs that "effectively modify social and cultural beliefs may be particularly promising in encouraging girls to pursue STEM areas for study and careers" (p. 4). Durlak, Weissber, and Pachan (2010) also note the importance of promoting personal and social skills as part of out-of-school-time (OST) program efforts. In particular, they advocate for skills designed to enhance self-efficacy and self-esteem.

The free OST program described in this chapter, Las Chicas de Matemáticas, incorporates many of the recommended research-based practices regarding disposition and motivation toward learning mathematics, as well as a learning environment that inspires women to pursue STEM-related careers and incorporates characteristics of successful outreach programs. Details are also provided on how this program integrates inquiry-based and hands-on learning, offers female role models in STEM-related careers, and promotes personal and social skills.

PROGRAM DESCRIPTION

Before introducing the camp at the University of Northern Colorado (UNC), the project director (PD) and a colleague implemented a scaled-down version of Las Chicas de Matemáticas (http://www.unco.edu/nhs/mathsci/mathcamp/) at the PD's previous institution. In 2008, the PD and two other mathematics faculty collaborated to conduct the first camp at UNC. The camp was offered each year from 2008 to 2014, except 2011 due to a lack of funding, and it will be offered again in the future as funding allows. The purpose of this free residential camp is to introduce 32 Colorado high school girls to the independence of college life, collegiate mathematics, STEM careers, and female role models who are passionate about mathematics. In this section, I describe the camp, the staff and their roles, and challenges that arise with this camp.

Camp Details

After securing funds to offer the camp free of charge to all participants, advertising for the camp commences. In February before each summer camp, the PD advertises the camp via past camp participants and through a Colorado mathematics email list that reaches K–12 teachers, academic counselors, and college professors who have signed up to be on the listserv. It is also advertised through the National Girls Collaborative Project (www.ngcproject.org). The application deadline is mid-March, and applicants are notified about acceptance decisions by the first of May. The time gap is needed for handling the large number of applicants, approximately 200 per year.

Any young woman who has successfully completed a course equivalent to the first year of algebra and who will enter grades 9–12 in the upcoming academic year is eligible to apply. Applicants must complete and submit the application form found at the Las Chicas de Matemáticas website, along with a statement indicating why they are interested in attending the camp. They must also include a letter of recommendation from a current mathematics teacher who can attest to the applicant's success in mathematics and other personal attributes such as persistence, level of engagement in mathematics courses, leadership, an ability to work well with others, and so forth. In selecting the final 32 participants, the PD seeks diversity in age, school size, residence location in Colorado, and race/ethnicity based on surname. Although this diversity is important, the applicant's letter of intent is instrumental in the final decision. Specifically, the PD strives to select applicants who want to be the first in their family to attend college, who speak passionately about mathematics, or who have minimal opportunities for such an experience due to geographic isolation. The applicants speak passionately about mathematics through comments that indicate math is their favorite subject, that they did not like math until a teacher encouraged them, that math is like a fun puzzle, or "I want to be a mathmagician."

Selecting the camp participants is challenging and is further complicated by the fact that past participants re-apply. Several participants attended the camp three years in a row and one would have attended four years had a 2015 camp been held. Having return participants is advantageous because they help newcomers adjust to the busy schedule, make friends, and overcome homesickness. The past participants are especially supportive in assuring the newcomers that they will be able to do the college-level mathematics. In an effort to give other young women an opportunity to participate in the camp, the number of returning campers is limited to eight participants per year. The eight are chosen based on how well they adapted to the challenging mathematics, got along with the other participants, and developed into leaders. The PD attempts to distribute returning participants among the grades 10–12 applicants.

The camp generally occurs during the second week of June, beginning on a Sunday afternoon and ending on Friday evening. Table 1.1 contains a typical camp schedule, which I discuss in this section. The journaling aspect of the camp is discussed in the Program Effectiveness section. After checking into the dormitory on early Sunday afternoon, the participants attend a parent/chica orientation with all camp staff. During this time a few returning chicas take time to share their past experiences as a camp participant. Over the years, the orientation has been followed by various activities. Initially, there was a picnic for the participants, parents, and staff, but this did not result in interaction among the chicas and it was expensive. Thus, in the most recent years the parents leave after the orientation and the chicas engage in icebreaker activities. One year the chicas worked in teams of four to participate in a scavenger hunt where they solved mathematical tasks/puzzles that allowed them to advance in the scavenger hunt. The

TABLE 1.1 Las Chicas de Matemáticas Schedule

Time	Sunday	Monday	Tuesday	Wednesday	Thursday	Friday
7:00–8:30		Breakfast	Breakfast	Breakfast	Breakfast	Breakfast
8:30–11:00		AM Math	AM Math	AM Math	AM Math	Banquet Prep
11:00–12:00		Speaker	Speaker	Speaker	Speaker	Speaker
12:00–1:00		Lunch	Lunch	Lunch	Lunch	Lunch
1:00–3:30	Check-In & Orientation	PM Math	PM Math	PM Math	PM Math	Banquet Prep
3:30–5:00	Ice Breaker & Team Building	Study Time	Study Time	Study Time	Study Time	Check-Out
5:00–6:00		Dinner	Dinner	Dinner	Dinner	Banquet
6:00–8:00	Pizza & Ice-cream Social	Wall Climbing	Swim Night	Movie	Play	
8:00–10:00	Unpack & Journal	Study Time	Study Time	Study Time	Study Time	
10:00–11:00		Journal Time	Journal Time	Journal Time	Journal Time	

tasks included building three-dimensional solids with cubes after examining a two-dimensional perspective, creating a given picture with tangrams, and transforming graphs of functions using feather boas.

Other years, the chicas designed the camp t-shirt as part of the icebreaker activity. Heidi Olinger, CEO of Pretty Brainy (http://www.prettybrainy.com), facilitated this activity and discussed the role of science and mathematics in fashion design. Each team of four created and presented a product. At the end of the session, the chicas voted on their favorite design and the t-shirts were available by the end of the camp. Although this activity was very popular, it resulted in hurt feelings when certain designs were not selected. The hurt feelings and other tensions between different ethnic groups prompted the PD to invite RaKissa Cribari, a mathematics educator from Colorado University at Denver and a counselor for Colorado Youth at Risk, to facilitate the icebreaker session in 2014. She kicked off the camp by engaging the participants not only in icebreaker activities but also team-building activities where the chicas as a whole or in teams of four collaborated on fun problem-solving activities. The session resulted in lessons regarding being patient, including others, making a plan, starting over, asking questions, and so forth, behaviors and actions that served the chicas well throughout the week. Each year Sunday evening concludes with a pizza and ice cream party and a final interpersonal-connections activity where each participant and her roommate state something they have in common and something that is different.

The schedule for Monday through Thursday consists of a hearty breakfast, a two-and-one-half-hour mathematics session, a one-hour presentation by a female guest speaker, lunch with the guest speaker, two-and-one-half hours of a different mathematics session, study time, dinner, social activity, evening study time, and journal-writing time. This schedule dedicates 8.5 hours per day to mathematics. Throughout the week, the chicas work in two different teams of four, one in the morning and one in the afternoon mathematics session, but each chica always works with her roommate. This ensures that they can readily continue their work in the dorms. The PD creates the teams of four to include as much diversity as possible, but with some familiarity. For example, each team generally consists of a freshman, sophomore, junior, and senior, where the freshman and sophomore are roommates and the junior and senior are roommates. Another example combination might be two freshmen, a junior, and a senior. The PD also attempts to assign roommates such that one girl is from a rural school and the other from an urban school. Girls from the same school are not assigned to the same team because it is important for the girls to learn to work with different people. This team diversity also ensures different mathematical backgrounds, which contributes to the team effort on mathematical tasks. The girls may work with whomever they want during the study sessions. It is not uncommon for two teams to merge during the afternoon study session. The evening study session takes place in the dormitory. This structure allows the participants to interact with different groups who offer multiple perspectives while learning new mathematics.

Both mathematics sessions focus on college-level mathematics topics found in junior- or senior-level mathematics courses. In past years, the participants have studied mathematical modeling, number theory, recreational mathematics, logical reasoning involving visual perception (using the card game SET), polynomial fitting, non-Euclidean geometry, knot theory, geometry of complex numbers, mathematical structures, visual abstract algebra, higher geometry, and transformations on various surfaces. These novel topics allow the participants to problem solve and to connect algebraic and geometric reasoning. Further, each course is taught using inquiry-based learning where the chicas explore mathematics concepts through student-centered activities, student presentations, and technology such as *Excel* and *Geometer's Sketchpad*. In general, the faculty spend about 10–15 minutes introducing the new material at the beginning of the class period, and the participants spend the remainder of the session working and struggling together. Faculty bring the class together to either answer a question that everyone has or to have the chicas present some of their work. The class presentations are an opportunity for the participants to see various solutions to the same task and to pose questions to participants who are not in their groups. The chicas work with each other closely throughout the day, but they also have opportunities to interact with the daily speakers.

Besides offering a description of their day-to-day job, the daily speakers share their stories about preparing and paying for college, choosing a college major, changing careers, balancing career and family, and anything else they believe might be valuable to the chicas. In particular, they emphasize the importance of mathematics in their careers and how mathematics opened doors for them in terms of selecting and changing careers. Past speakers held such job positions as engineers (aerospace, chemical, civil, software, systems, and water resource), statisticians, research physicians, chemists, cryptographers, biologists, mathematicians, atmospheric scientists, chief executive officers, insurance agents, veterinary epidemiologists, architects, bio-chemists, computer scientists, physicists, and a land technician. These women usually work in industry or in academia at UNC, Colorado State University, or the Air Force Academy. Sample companies who have provided speakers include Wolf Robotics, National Jewish Hospital, National Security Agency, SOARS, Raytheon, State Farm, OtterBox, Noble Energy, CH2M Hill, and Rocky Mountain Instrumental Laboratories. The chicas have been fortunate to interact with high-profile women who are Caucasian, Hispanic, Asian, and African American. This racial/ethnic diversity allows participants to identify with the speakers and possibly see themselves in similar careers.

It is noteworthy that although each speaker is offered a small stipend, no one has accepted it. All the speakers have been gracious with their time and truly enjoy interacting with the camp participants. Chicas who are interested in a particular career have opportunities to interact further with the speaker during lunch. Subsequent to the camp, some chicas even shadowed some of the speakers on the job to learn more about given careers. The fact that the speakers are willing to be shadowed illustrates the speakers' commitment to serve as role models for the chicas.

Although the camp devotes a great deal of time to mathematics, time is also allocated for social activities. The evening social activities give participants an opportunity to interact with one another outside of the classroom. The social activities include rock climbing and other indoor gym activities, swimming, watching a movie, and attending a play. For many of the girls, it is their first time rock climbing or attending a play. The camp culminates with a banquet, where groups of four chicas present some of the mathematics they learned during the week. The chicas spend most of Friday preparing and practicing their PowerPoint presentations, but they get a brief break during the speaker time. Participants, their parents, their mathematics teachers, camp sponsors, and UNC dignitaries attend the banquet. During the banquet, we show a slideshow depicting the work and leisure activities that took place during the week and that contributed to the participants' mathematical, personal, and social skills.

In an effort to maintain participants' confidence level developed during the camp (as documented in the Program Effectiveness section), the participants return on a Saturday in early fall to share how the school year is going, to communicate what they have been able to take back to their mathematics classroom, and to create the webpage for that year's camp. Although not all of the young women are able to attend the one-day event, at least half of the participants return for the reunion with other chicas and the camp staff.

Camp Staff

The camp staff consists of the PD (a mathematics educator), a mathematician, and four counselors. The PD organizes the camp with assistance from the university's conference services, whose staff coordinates meals, social activities, and housing. The PD also teaches the afternoon course, and a mathematician teaches the morning course. This is always a Hispanic male, who serves as a valuable role model because male minorities are also underrepresented in STEM fields. Four female UNC undergraduate mathematics majors serve as camp counselors. The PD invites the counselors based on her classroom interactions with them or recommendations from other mathematics faculty. The counselors stay in the dormitory with participants, help in the classroom, assist with study sessions, and supervise the participants in general. Each counselor is responsible for two teams of four participants, who work together in instructional sessions. Sometimes the counselors are learning the mathematics content for the first time alongside the chicas. While this might sound problematic, the chicas appreciate that they are learning something the counselors do not know, which confirms that they are learning college-level mathematics. In an effort to expose the counselors to classroom material prior to the camp, the staff have access to an online file-sharing folder. The folder is also used to save student work and photos used for the banquet slideshow.

Besides the staff described above, every year there are a number of women who want to learn about the camp. Volunteers have included female graduate students from UNC and San Diego State University, a mathematics educator from UNC, mathematicians in and out of academia, a physicist, and a civil engineer who was a past speaker. The volunteers participate in all the activities, and their assistance in the classroom is quite valuable. Further, they serve as additional role models beyond the scheduled speakers.

Challenges

Offering the camp is not without challenges. The biggest is raising approximately $35,000 to cover costs for the summer week and follow-up activity in time to start advertising the camp in February. Fortunately, the UNC Development Office staff has worked closely with the PD to identify and secure funds. As a result of this collaboration, State Farm, Noble Energy, OtterBox, CH2M Hill, and the Women's Foundation of Colorado have sponsored the camp over the years. All of these organizations have also provided the camp with speakers. Other sponsors have included private and anonymous donors, the Mathematical Association of America/Tensor Foundation, and the host institution. The UNC School of Mathematical Sciences, College of Natural and Health Sciences, Office of the Provost, and Office of Enrollment Management have all funded the program in various ways. This includes sponsoring the banquet, donating classroom space, and paying for classroom materials. The UNC Noyce Scholars Program has also sponsored the program by providing support for some of the counselors.

Other hurdles that are more difficult to tackle include some applicants' lack of commitment to the program and some participants' lack of respect for other cultures. For example, it is common for some applicants to not show up for the camp, which can be very frustrating. Preparing a list of alternates who live close to campus has alleviated this challenge, but it is time-consuming for the PD to make adjustments at the last minute. In 2014, the PD included participants' teachers on all email correspondence and for the first time, all selected participants showed up at camp. In fact, some teachers drove participants to camp and stayed for the orientation. A second challenge is that, inevitably, one or more participants do not want to be at camp because of the cultural diversity, as indicated in their journals (discussed below), and they thus behave disrespectfully toward other participants. It is these circumstances that inspired team-building activities as part of the icebreaker session for the most recent camp. The fact that there was no tension due to participant cultural background during that camp might indicate that such team building is important for similar programs. Despite experiencing some challenges, Las Chicas de Matemáticas appears to be an effective OST program based on the results shown by data collected during the camp.

PROGRAM EFFECTIVENESS

Every year the PD strives to select half the participants from rural areas and the other half from urban parts of Colorado. Figure 1.1 illustrates the hometowns of the young women who participated in the camp during

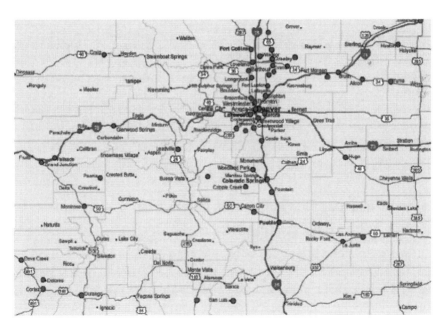

Figure 1.1 At least one chica has attended this program from each location indicated by the larger dots on the map.

2008–2010 and 2012–2014. In 2008, the average camper age was 16 years; in all the other years the average age was 15 years. These demographics are not surprising because a majority of the applicants are entering sophomores or juniors. Given that applicants do not document their race/ethnicity on the application form, this information is not inferred until the participants arrive on campus. Based on surnames and conversations with the young women, I categorized the participants as Hispanic, Caucasian, and "Other." The "Other" category consists of Indian, African American, Asian, Iranian, and Turkish. Although this approach might be considered a limitation of this paper, I feel confident in the accuracy of the reported percentages because I develop a strong rapport with the chicas and they tend to volunteer this information. In 2008, the race/ethnicity breakdown was 60% Hispanic, 33% Caucasian, and 7% Other; in 2009 it was 43% Hispanic, 40% Caucasian, and 17% Other; in 2010 it was 33% Hispanic, 57% Caucasian, and 10% Other; in 2012 it was 50% Hispanic, 40% Caucasian, and 10% Other; in 2013 it was 50% Hispanic, 40% Caucasian, and 10% Other; and in 2014 it was 56% Hispanic, 34% Caucasian, and 10% Other. The fact that the camp attracts young women from diverse cultures and diverse parts of Colorado is one measure of program effectiveness. Program success is also measured by data collected via a Likert-scale attitude survey and participant journals.

Attitude Survey Results

In 2012 the PD started to assess the camp's influence on participants' confidence level and attitude toward mathematics using a Likert-scale pre- and post-test attitude survey (see Appendix A). This instrument was adapted from Chacon and Soto-Johnson's (2003) attitude survey. Each item is scored from 1 to 5, ranging from 1 for strongly disagree to 5 for strongly agree. Items 1, 3, 6, 9, 10, 13, 17, and 19 are reversed; thus, those statements are scored with 1 for strongly agree to 5 for strongly disagree. The participants complete the attitude survey Sunday evening and Friday afternoon. Each participant receives a total score for both the pre-test and the post-test attitude survey, and the post-test minus pre-test attitude score is computed for each participant. A paired-samples *t*-test is used to compare mean scores of the two tests for potentially significant differences. According to the Central Limit Theorem, a sample size of at least 30 is sufficient to achieve normality; thus, the *t*-test is appropriate (Agresti & Finlay, 2009). The *p*-values, shown in Table 1.2, indicate that participants' dispositions toward mathematics (attitudes, confidence) are significantly different from the first to the last day of camp. This transformation is further substantiated in participants' journal comments.

	TABLE 1.2 Participant Attitudes at Camp Entry and Exit		
Year	Post-Minus Pre-Test Mean Score	t-Score	p-Value
2012	2.68 (N = 30)	2.46	<.01
2013	2.74 (N = 31)	2.75	<.01
2014	2.74 (N = 31)	2.75	<.01

Note: Not all sample sizes are 32 because some participants did not complete the final survey or the camp.

Journal Results

Daily journal writing has been a camp component since the program was founded in 1999. The primary purpose of the journals is to serve as a mechanism for safe communication between the PD and the participants. Sunday evening the participants receive a journal, and every evening the PD provides the participants with journal prompts. During the morning mathematics session, the PD reads all the journals and responds to participant comments, concerns, and questions. This is an efficient way to communicate with each chica and to assess how the chicas are responding to the camp activities on a daily basis. Sometimes journal comments alert the PD to potential roommate or classroom conflicts, which the PD addresses by speaking directly to the chica(s) or by making the counselors aware of given situations. A counselor photocopies all journal entries, and the journals are returned to the chicas during the afternoon study session. Sample prompts from 2014 appear in Appendix B. The Sunday evening journal prompts have been the same since the inception of the camp. The Thursday evening prompts have been similar since the 2012 camp. Question Four was added for the 2014 camp in an effort to determine the influence of the team-building and icebreaker session.

In addition to serving as a communication tool, the journals function as an instrument to help evaluate the effectiveness of the program. In an effort to synthesize the journal entries within categories, an Excel spreadsheet was created (see Appendix C). The column headings of "worry," "nervous," "confidence," "mathematics," "community," "career," "high school experience," "different ways of learning," and "general" are based on the journal prompts and themes that appear in the chicas' comments. Documenting this information for camp years and days of the week provides insight into trends and changes over this time, especially for return participants. These qualitative data support the quantitative finding that participants' confidence and attitude toward mathematics improve during the week, which many participants attribute to the IBL classes. Evidence also suggests that the chicas' social skills develop over the camp week and that the speakers spark awareness of STEM careers. The results presented here are from the 2012–2014 camp due to the more consistent nature of the prompts used during that time frame. The sample quotes presented were selected because they are representative of other chicas' comments. (All names are pseudonyms.)

On the first day of camp the participants indicate that they are most worried and/or nervous about not being smart enough, not making friends, and being away from home for the first time. After the parent/chica orientation and the icebreaker activities, the chicas are more cognizant that the mathematics is college-level, that some participants have completed AP calculus, and that some of the chicas have attended similar programs for accelerated students. This awareness might contribute to participant comments shown below. Every year some chicas, such as Lucille, feel confident in their abilities but are concerned about being able to grasp the material quickly. Some, like Teresa and Belinda, immediately compare themselves to the other chicas and fear they are not smart enough. Others, like Leslie and Anna, worry about what others might think of them if they make a public mistake or give up on a task.

> **Lucille (2012):** I am most worried about classes . . . because I am used to classes where things come naturally or very easily to me.
>
> **Meg (2012):** I'm a little worried about the math, that I won't understand or that it will take me a long time to learn it.
>
> **Leslie (2013):** I'm personally very worried that I will not be able to do the math that will be required. I'm scared that I will make a "silly" mistake in front of everyone. I'm most terrified of being unable to grasp the topics we will learn about.
>
> **Teresa (2013):** I am most worried about whether or not my math skills will compare to the other girls' because there are some very talented girls here.
>
> **Anna (2014):** I am worried about not having enough "grit" or perseverance to solve a very challenging question(s) because I don't want to be known as a quitter.
>
> **Belinda (2014):** I am most worried about being behind all the girls in math since I've only taken 1 high school math class. I've never taken geometry. I'm nervous about being able to do all the problems that I am given.
>
> **Melissa (2014):** I am worried that I will struggle with the new math because it might not build on concepts that I already know.

Fortunately, for most, these concerns are usually put to rest by Monday evening, as documented below. Realizing that the content was new to everyone, that the staff believed in them, that they could ask questions to their group members and counselors, and that they could be creative with their work appeared to ease participants' concerns regarding the difficulty of the mathematics content. Many, such as Belinda, begin to indicate an increase in confidence by Wednesday evening and certainly by Thursday evening.

Lucille (2012): My worries about classes were tamed pretty instantly once we started . . . I'm not sure how it happened exactly but I think my brain knew I would be okay and so did my nerves. (Monday)

Leslie (2013): Honestly, the best part of camp so far has been how the class has been taught. I didn't feel like the class was held back or talked down to. We were treated as if we could do the math and because of that, we were able to excel. There was no doubt in our ability. I didn't even think of doubting my own. (Monday)

Teresa (2013): My biggest worry from yesterday is completely gone now. I realized that I'm not the only one who struggles and that I am just as capable at solving problems as others. (Monday)

Melissa (2014): I started to tame my nerves a bit because I asked them [group members] questions about how/why they did something and I also asked the counselors a few questions. (Monday)

Belinda (2014): I used to think the mathematics was an impossible task but now they're doable (yet still challenging) and fun! I'm learning to talk more and be confident in my abilities. (Wednesday)

Meg (2012): As we began to understand the concepts it got easier. I can play around with anything that I'm curious about. The questions and assignments really encourage creativity and curiosity. (Thursday)

Anna (2014): I believe that my attitude towards math has become more persistent because eventually the problem can be solved. (Wednesday). In such a short period of time I feel more confident about math because it can be difficult but it can also be solved. (Thursday)

Although eventually there is a change in participants' disposition, the change is not always easy. For example, in 2014 Julieta was extremely overwhelmed on Monday and wanted to go home. Monday night she wrote: "The mathematics was very difficult! I've never seen it before. . . . I thought I was going to be introduced to geometry and then boom! Do this, that, this, that, this, this and this! Woah I didn't know what to do." After reading this, the PD asked her counselor to spend more time with Julieta and to ensure that Julieta's group was working collaboratively. Tuesday night Julieta wrote in her journal that the "best part of the day was being able to do the math and getting it right." She elaborated that she "felt accomplished" and understood the math because she "tried." This attitude continued into her Wednesday night journal where she explained, "My attitude has changed towards the math because I actually give it a try now. The first day I was like whatever, now I actually try." Thursday she concluded with, "Doing the math felt groovy awesome!" Like Julieta, on Tuesday night many of the chicas make comments such as "BAM—I got it!," "the mathematics clicked," "I had an ah-ha moment," and "I connected the content."

It can be difficult to identify what influences participants' change in confidence, but the IBL classes seem to be a contributing factor. As documented below, this teaching style allows the participants to recognize that mathematics goes beyond textbooks, that making mistakes is okay, and that collaborating is powerful, efficient, and fun. It also allows them to be creative, as mentioned above by Meg (2012) and below by Leslie (2013). All of these lessons are important in and out of the mathematics classroom. Selected participant comments include:

Daisy (2012): This camp has showed me that math isn't just in a textbook. I LOVE HOW YOU CELEBRATE OUR MISTAKES. It was so cool to see girls go up to the front and you have us clap for them even when they were wrong.

Gabby (2012): I am used to being the only one working when doing group work at school. Here I find the extra brain power helpful!

Leslie (2013): I think this is the best way to learn math. In a "normal" classroom all of the creativity has been sucked out of math.

Belinda (2014): I'm learning I prefer working in groups compared to individually. It's more fun and helps learn the concepts better. It's weird to be encouraged to talk more since most schools require you to work silently and alone.

Sarah (2014): By working together there is always a way that we find to make solving problems less of a chore and more entertaining.... We power through them, stay persistent, and the feeling of accomplishment at the end is so rewarding.

The data also indicate that transformed confidence continues into the next year for some participants. For example, at the end of the 2012 camp, Marisa wrote: "I found that the beginning of the camp was super hard for me but after a few days I start[ed] to get the math problems and I just love doing it. I've learned so much and this camp is helpful [for] me." On Sunday night of the 2013 camp, Marisa wrote: "I'm not worried about anything and I'm really excited about tomorrow. I want to challenge myself to do math." Similarly, on Wednesday night of the 2013 camp, Sarah commented:

It seems surreal to think that I am learning the same material that juniors in college are learning and that makes me feel great about myself and gives me a real feeling of accomplishment. Although the problems are challenging, that just assures me that I'm growing in the mathematics and we are learning, and to keep persevering to overcome the challenges.

The following year, she started her journal by exclaiming, "This is where I belong." Such declarations seem to suggest that the positive change in disposition and confidence extends beyond the camp week.

The second biggest concern that the chicas report pertains to making friends, but this again diminishes quickly. Some chicas break free of their shyness after the rock climbing and gym evening social event. For others, collaborating and struggling together in the classroom is a source of bonding. Some, such as Diana, a 2013 camp participant, realize how important and easy it can be to make friends. On Sunday evening, Diana wrote, "The thing I am most worried about is being social. I'm not typically very good at this...I am very shy." Monday night she commented, "As it turned out it was really easy to make friends." By Tuesday night Diana was most proud of her ability to make friends and wrote, "This early in the week I feel I've already accomplished something I have never ever been able to do. I was able to be friends and talk to everyone and continue on conversations with them." On Wednesday evening Diana realized how much she wanted and needed to be more social:

> This week I learned something really weird that I've always thought to be wrong about me. It's that no matter how much I can go without socializing and human contact I am still human....I need to be around people....I never realized this about me.

The third biggest concern regards being away from home for the first time. This is particularly true for the Hispanic participants. For example, on Sunday night, Gabby, a 2012 participant, wrote that she was most worried about "being a significant distance away from my parents." On Thursday evening, she reported, "I thought that I wouldn't be able to wait to get home. It turned out to be the exact opposite." Other chicas with the same initial concerns wrote comments such as "fastest week ever," "wish I could stay here forever," or "found a new family in the friends I've made" as part of their last journal entry.

Although most participants resolve their concerns, some do not and leave the camp. For example, one young woman who was worried that she was not smart enough and was concerned about making friends did not finish the program. In her journal, she admitted that she had no interest in mathematics and only applied to the program to strengthen her college applications. Another participant left the camp because she felt uncomfortable with the cultural diversity. Fortunately, this is uncommon and most campers come to appreciate the diversity of the program participants. For example, Esperanza, a 2014 chica, expressed, "I think the quasi targeting of Hispanics is kind of weird because firstly not all of the campers are Hispanic and also it creates a weird cultural/background/race emphasis which can just get awkward." At the end of the camp she remarked that although she initially had "some major concerns about the awkward cultural differences," the camp exceeded her "expectations in the best possible way." Most participants share Sarah's sentiments, which she noted on the last day of her second year of camp:

> The diversity of the girls at camp has always been something that I don't really notice. After getting to know them, I really just think of the words, the actions, and the memories that we all shared. It was cool though to see how we all share the common interest of math, even though we all have different backgrounds.

A primary purpose of the camp is to expose the participants to STEM careers. Although many of the participants arrive at the camp knowing that they love mathematics, few are aware of what they can do with a strong mathematics background, and many lack confidence in their ability to do college-level mathematics. All of the participants find the speakers "inspiring" and appreciate how the speakers "open [their] eyes to the possible career choices." It seems clear that the speakers serve as role models and as resources for the chicas. For example, in 2014, Gretchen shadowed the architect that inspired her quote shown below. Some speakers even influence the chicas enough for them to consider majoring in mathematics. Such is the case with Tiana, who made the comment shown below in her third year of camp. The fact that it took Tiana three years of camp to consider mathematics as a major might indicate that young women need multiple and repeated exposures to possible STEM careers.

Meg (2012): I didn't even know epidemiology was a field. I liked the speakers, it was one of my favorite parts of the day just to hear their stories and to see how they followed their dreams and overcame adversary [adversity] was really inspiring.

Gretchen (2012): I am definitely now considering architecture as a future career option.

Anna (2014): I loved the fact that the speakers were real women who had achieved a title in math/science careers some even in male-dominated fields like engineering. It would be an understatement to say it was inspiring.

Tiana (2014): Today while listening to the speaker I had a weird epiphany; I don't want to do theatre anymore. I realized that isn't something I will pursue after high school and this year I want to focus more on my school work.... I feel like I should utilize my potential in other areas as well. So I have come to the conclusion as I am writing this... that I can double major in something music-y and something math-y.... So today/yesterday have been a little weird for me. Weird but good.

PRACTICAL IMPLICATIONS

A question frequently posed regarding Las Chicas de Matemáticas is whether the participants pursue STEM careers as a result of participating in the program. This is a difficult question to answer because this requires a control group and a longitudinal study, which would be costly and time consuming. Further, many confounding variables, such as home life, participation in extracurricular school activities, socioeconomic status, parents' educational background, and so forth influence young women's choices. As Laursen et al. (2007) argue, "When the intervention strategy is short in duration . . . the immediate outcomes of such events are primarily affective" (p. 50). Thus, it is possible that the most we can hope for is that providing learning environments where the young women experience success with college-level mathematics, that sparking the young women's interest and enthusiasm for mathematics, and that exposing them to women role models with STEM-related careers will translate to pursuing careers in STEM-related fields.

Although this research does not answer the question as to whether the participants pursue STEM careers, the data suggest that the program situates the participants to be successful in a STEM career, if they choose to pursue such a field. The quantitative data illustrate that the participants' confidence in their abilities and their disposition toward mathematics improved from the beginning to the end of the camp. This result might seem surprising in that most participants apply to the camp because they already enjoy their mathematics courses and perform well in their classes. The statistically significant improvements in disposition could be due to participants' exposure to college-level mathematics and the girls' initial uncertainty about their ability in relation to such content. In any case, the fact that statistically significant favorable differences consistently appear is promising. This might imply that the participants leave the camp feeling even more confident in their ability to do mathematics, which could continue through the remainder of their high school and perhaps college years.

The qualitative data support the quantitative data and findings regarding the benefits of IBL for women (Eccles, 1989; Jobe, 2002/2003; Laursen et al., 2007; Laursen et al., 2014). This atmosphere allowed the participants to struggle and succeed in a safe environment within their established instructional groups. Although the literature suggests that parents tend to attribute their son's mathematics success to natural talent and their daughter's to effort (Aschbacker et al., 2010; Eccles, 1989), the camp staff stresses the importance of both. Such a message is essential because "most child prodigies are highly successful—but most highly successful people weren't child prodigies" (Ellenberg, 2014, p. C3). The camp participants are routinely reminded of the importance of developing their intelligence, working hard, and making mistakes. In their journals many participants remarked that they "enjoyed the challenging material," appreciated that it was "ok to make a mistake," and knew that they would be able to do it because everyone "believed they could do it." Such comments allude to the importance of challenging young women and providing them with situations where they struggle and succeed due to persistence, collaboration, and hard work. As teachers, we must promote the notion that making mistakes is part of learning. Convincing students of this might be easier when there is an established student–teacher rapport.

The participants also commented that they valued how the structure of the instructional sessions encouraged creativity. Supporting creativity as a feature of IBL courses is not well documented in the literature, but Jolly (2009) affirms that giftedness is a measure of both intelligence and creativity. The speakers also reinforced the notion of creativity. For example, the architect showed pictures of some of her home and office designs that mimicked the natural surroundings of those structures. Several of the engineers also noted that they enjoyed their jobs because they had opportunities to integrate their creative and scientific personalities. Certainly, Heidi Olinger, who facilitated the t-shirt design activity, engaged the chicas in a creative yet mathematical task. Given that participants were attuned to the fact that they were allowed and encouraged to be creative, it might be important to activate young women's creative thinking in STEM-related courses and extracurricular activities. It also seems crucial that young women be aware that STEM-related careers can be creative.

Many are trying to improve the low proportion of women in STEM-related careers, but in spite of programs designed to increase these numbers, the United States has not made significant gains. Marra, Peterson, and Britsch (2008) acknowledge that from the perspective of equal opportunity, the number of women who pursue STEM careers is too low. They also comment that in order to meet the projected needs of STEM professionals, we as a community must step forward to "extend the capacity, impact, and sustainability of existing and evolving girl-serving STEM projects and programs" (p. 120). Las Chicas de Matemáticas is such an effort, but the PD acknowledges that this is a small-scale program. The participants, teachers, and parents all request that the camp be longer and designed for more participants. Although this might sound ideal, it would require more funds and more time from the staff. Further, the intimate size allows the staff to dedicate a great amount of time to each camp participant. The 1:4 ratio of staff to chicas allows the staff to address the individual, social, and academic needs of each chica. This could be challenging to accomplish with a bigger program. On the other hand, the staff is open to inviting teachers to serve as apprentices to the faculty so that they may observe and participate in IBL teaching methods. This would also give teachers a chance to witness how the faculty challenge but encourage the young women in such a way that instills a desire to learn college-level mathematics, the gateway to all STEM fields. Researchers might want to explore how such an experience influences the teaching of experienced mathematics teachers.

CLOSING COMMENTS

Two aspects of the camp contribute to its uniqueness. The first is that the camp is free and thus attracts young women who do not have the financial resources to attend similar programs. Accordingly, this structure affords young women with mathematics potential but without financial resources an opportunity to expand their knowledge of mathematics, to engage with female STEM role models, and to consider careers in STEM-related fields. The second component is the staff's dedication to the program, which is grounded in their own educational and personal experiences. In fact, the PD intentionally created a Spanish name for the camp in an effort to connect to Hispanic parents who do not tend to allow their daughters to participate in such experiences. Her hope was that this would transcend the language barrier with the Hispanic parents. The PD's background is similar to some participants' backgrounds; she can thus identify with the participants. This staff-participant identification allows the chicas to see what is possible and to dream about reinventing their future. Researchers might want to investigate how participants with backgrounds similar to that of the staff respond to the camp activities compared to participants with differing backgrounds. Such investigations could contribute to research on the importance of STEM role models.

Offering OST programs, such as Las Chicas de Matemáticas, is time consuming and can be emotionally draining for the staff. However, it is extremely rewarding and beneficial for everyone involved in these programs.

APPENDIX A
Las Chicas de Matemáticas
Attitude Survey

Name: _____   Age: _____

Please put a check (√) in the box that most indicates your reaction to each of the following statements.

Statement	Strongly Disagree	Disagree	Neutral	Agree	Strongly Agree
1. I worry that math classes are difficult for me.					
2. I keep trying in math, even if the work is hard for me to do.					
3. I get very mad when I have to do my math homework.					
4. I look forward to my math classes.					
5. I am certain I will be good in math when I graduate from high school.					
6. I get nervous doing math problems.					
7. I get good grades in math.					
8. I work as hard as I can in math.					
9. I worry that I will get poor grades in math.					
10. I am not good in math.					
11. I enjoy math.					
12. I am interested in the things I learn in math.					
13. I only do math because I have to.					
14. Math problems can be solved in different ways.					
15. I think about how the math I have learned can be used in everyday life.					
16. I try to understand how the math I learn is used in other subjects.					
17. To learn math, I try to memorize every step in a solution.					
18. I enjoy solving math problems.					
19. I get upset when I can't solve a math problem.					
20. I like to do math.					

APPENDIX B
Sample Journal Prompts

Sunday Evening: Answer each of the following questions in your journal. Please use complete sentences and complete thoughts.

1. What has been the best part about camp so far? Why?
2. What are you most worried about regarding the camp? Why?
3. What are you most excited about regarding the camp? Why?
4. What are you most nervous about regarding the camp? Why?
5. In what ways will you be able to contribute to your groups?
6. Please share anything else that you would like.

Tuesday Evening: Answer each of the following questions in your journal. Please use complete sentences.

1. What was challenging about the mathematics today? Why? How did you defeat this challenge?
2. What was the best part of the day? Why?
3. What was the most difficult aspect of the day? Why?
4. Describe how the teaching of mathematics at math camp is the same and different from the teaching of mathematics at your school. Describe how you are adjusting.
5. Describe how you have implemented suggestions from Dr. Cribari (Sunday night's speaker) as you work together.
6. Lucia Flores talked about preparing for college. Tell me what you learned (i.e., you didn't already know it) and how that was helpful. If you knew it all, tell me about something that you know but that she didn't share.
7. Please share anything else that is on your mind.

Thursday Evening: Answer each of the following questions in your journal. Please use complete sentences and complete thoughts.

1. Please share your reactions to our speakers this week. What did you like or dislike about them?
2. Please share your reactions to the social activities this week. What did you like or dislike about them?
3. Please share your reactions to the mathematics that you have learned this week. What did you like or dislike about it?
4. Please share your reactions to the diversity of the young women who are at the camp.
5. Please share how you feel different from the beginning of the week. If you feel the same, tell us why that is the case.
6. Please share your overall experience with the camp. How can we improve the experience for future chicas?
7. Please share anything else that you would like.

APPENDIX C
Journal Analysis Chart

Name	Year	Day	Worry	Nervous	Confidence	Math	Community	Careers	High School Experience	Different Ways of Learning	General
	2012	Sun									
		Mon									
		Tues									
		Wed									
		Thurs									
	2013	Sun									
		Mon									
		Tues									
		Wed									
		Thurs									
	2014	Sun									
		Mon									
		Tues									
		Wed									
		Thurs									

REFERENCES

Aschbacher, P. R., Li, E., & Roth, E. J. (2010). Is science me? High school students' identities, participation, and aspirations in science, engineering, and medicine. *Journal of Research in Science Teaching, 47*(5), 564–582.

Agresti, A., & Finlay, M. H. (2009). *Statistical methods for the social sciences (4th ed.).* Upper Saddle River, NJ: Pearson.

Beede, D., Julian, T., Langdon, D., McKittrick, G., Khan, B., & Doms, M. (2011). *Women in STEM: A gender gap to innovation.* Washington, DC: U.S. Department of Commerce, Economics and Statistics Administration. Retrieved from http://www.esa.doc.gov/sites/default/files/reports/documents/womenin-stemagaptoinnovation8311.pdf

Beyers, J. (2011). Student dispositions with respect to mathematics: What current literature says. In D. J. Brahier & W. R. Speer (Eds.), *Motivation and disposition: Pathways to learning mathematics [73rd NCTM Yearbook]* (pp. 127–140). Reston, VA: NCTM.

Chacon, P., & Soto-Johnson, H. (2003). Encouraging young women to stay in the mathematics pipeline: Math camps for young women. *School Science and Mathematics, 103*(6), 274–284.

Commission on the Advancement of Women and Minorities in Science, Engineering, and Technology Development. (2000). *Land of plenty: Diversity as America's competitive edge in science, engineering and technology.* Arlington, VA: National Science Foundation. Retrieved from http://www.nsf.gov/pubs/2000/cawmset0409/cawmset_0409.pdf

Durlak, J. A., Weissberg, R. P., & Pachan, M. (2010). A meta-analysis of after-school programs that seek to promote personal and social skills in children and adolescents. *American Journal of Community Psychology, 45*(3–4), 294–309.

Eccles, J. S. (1989). Bringing young women to math and science. In M. Crawford & M. Gentry (Eds.), *Gender and thought: Psychological perspectives* (pp. 36–58). New York, NY: Springer-Verlag.

Ellenberg, J. (2014, May 30). The wrong way to treat child geniuses. *Wall Street Journal.* Retrieved from http://online.wsj.com/articles/the-wrong-way-to-treat-child-geniuses-1401484790

Executive Office of the President of the United States. (2011). *Women and girls in science, technology, engineering, and math (STEM).* Retrieved from http://m.whitehouse.gov/sites/default/files/microsites/ostp/ostp-women-girls-stem-november2011.pdf

Guynn, J. (2014, July 2). Sexism in tech buzz gets louder. *USA Today—Fort Collins, Coloradoan,* 4B–5B.

Jobe, D. A. (2002/2003). Helping girls succeed. *Educational Leadership, 60*(4), 64–66.

Jolly, J. L. (2009). The National Defense Education Act, current STEM initiative, and the gifted. *Gifted Child Today, 32*(2), 50–53.

Laursen, S., Hassi, M. L., Kogan, M., & Weston, T. J. (2014). Benefits for women and men of inquiry-based learning in college mathematics: A multi-institution study. *Journal for Research in Mathematics Education, 45*(4), 406–418.

Laursen, S., Liston, C., Thiry, H., & Graf, J. (2007). What good is a scientist in the classroom? Intervention in K–12 classrooms. *CBE Life Sciences Education, 6*(1), 49–64.

Marra, R. M., Peterson, K., & Britsch, B. (2008). Collaboration as a means to building capacity: Results and future directions of the National Girls Collaborative Project. *Journal of Women and Minorities in Science and Engineering, 14*(2), 119–140.

Mataric, M. J., Koenig, N., & Feil-Seifer, D. (2007, March). Materials for enabling hands-on robotics and STEM education. In *AAAI Spring Symposium on Robots and Robot Venues: Resources for AI Education.* Retrieved from http://www.academia.edu/2476297/Materials_for_Enabling_Hands-On _Robotics_and_STEM_Education

National Science Board. (2006). *Science and engineering indicators* (Appendix 4–33). Retrieved from http://www.nsf.gov/statistics/seind06/

Seymour, E. (2006, March 15). *Testimony offered by Elaine Seymour.* Washington, DC: Hearing on Undergraduate Science, Math and Engineering Education: What's Working?, Research Subcommittee of the Committee on Science of the U.S. House of Representatives.

Seymour, E., & Hewitt, N.M. (1997). *Talking about leaving: Why undergraduates leave the sciences.* Boulder, CO: Westview Press.

Soto-Johnson, H., Craviotto, C., & Parker, F. (2011). What motivates mathematically talented young women? In D. J. Brahier & W. R. Speer (Eds.), *Motivation and disposition: Pathways to learning mathematics* [73rd NCTM Yearbook] (pp. 127–140). Reston, VA: NCTM.

Weinberg, J. B., Pettibone, J. C., Thomas, S. L., Stephen, M. L., & Stein, C. (2007). The impact of robot projects on girls' attitudes toward science and engineering. Retrieved from http://www.cybercamp.ca/Benefits_of_Robotics_files/ girls2007.pdf

CHAPTER 2

RELATIONSHIPS AMONG GIRLS, CONTENT, AND PEDAGOGY IN THE MATHERSCIZE SUMMER MATHEMATICS CAMP

Diana B. Erchick

ABSTRACT

Matherscize is a summer mathematics day camp for girls entering grades 5, 6, or 7 the autumn following the camp. Campers are typically 10–14 years old. The program is offered free of charge at a regional campus of a large, mid-western research university with no requirements for particular camper skill or performance levels prior to attending the camp. Despite some girls' resistance toward the mathematics they encountered, findings suggest that a process approach favorably influenced the girls' mathematics understanding and challenged their relationship with mathematics. The pedagogy employed in Matherscize tended to improve the girls' dispositions toward mathematics but in somewhat troubling ways, as perceptions of mathematics that could be problematic for the girls in future endeavors also surfaced. Observations by instructors and researchers indicated that inter/intrapersonal relationships

Out-of-School-Time STEM Programs for Females, pages 29–53
Copyright © 2017 by Information Age Publishing
29

in Matherscize supported participant learning, and the girls valued those relationships even when they experienced conflict.

I developed Matherscize as a summer mathematics camp, with a vision toward understanding and supporting girls' experience with mathematics. Because I have always loved mathematics, I find it hard to understand why anyone would not want to learn more about mathematics, to understand more, to further develop a sense of wonder, or simply to enjoy the discipline. Still, as a woman in mathematics and now in mathematics education, as a student, and as a teacher, I am attuned to issues surrounding mathematics learning. I am intrigued by questions involving the role of the teacher and of gender, pedagogy, and the environment, as well as the structure of school mathematics content and shortcomings in the field of mathematics for many women. While such elements can contribute to women's positive experiences with mathematics, they can also become barriers to meaningful mathematics experiences. Together, the elements comprise what I call the Space/Barrier (Erchick, 1996), a terrain of elements educators and students work through or around during mathematics experiences. As mathematics educators, we work in ways intended to strengthen contributing elements and eliminate hindering elements, so that those encountering mathematics can learn, enjoy, and grow from the experience.

In my work I try to identify teaching and learning elements that present barriers to mathematics learning. From this, I hope we can learn to design more supportive environments. Matherscize is intended to be such an experience, and data collected on the program is intended to determine elements that make a difference in and for student learning.

BACKGROUND

An Office for Civil Rights (2012) report notes that even though boys still outnumber girls in many high school mathematics courses, the margin is small and girls are indeed nearly proportionately represented in rigorous high school mathematics courses. However, the report states:

> Despite women's gains in some nontraditional fields as a whole, the rate of female enrollment in certain career clusters remains at persistently low levels. In 2009–2010, females made up less than 25% of participants in science, technology, engineering, and math programs nationally (21% at the secondary level and 24% at the postsecondary level). (p. 3)

Many young women continue to opt out of fields that require strong use of mathematics, and they tend to fail to identify with or to see the usefulness of mathematics (Becker, 2002; Ceci, Ginther, Kahn, & Williams, 2015;

Ceci & Williams, 2010; Cheryan, 2012; Diekman, Brown, Johnson, & Clark, 2010; Good, Rattan, & Dweck, 2012; Herzig, 2002; Office for Civil Rights, 2012). Thus, even when expectations and policies support females in mathematics, the outcomes are not always favorable. What influences positive mathematics experiences for females is not always clear. I propose that we need to understand those factors as well as how they relate to each other. To do this, we can explore experiences in terms of learners' relationships with specific elements: mathematics content, pedagogical approaches, and interactions with others in the environment. This is a viable framework for conducting research in a girls' mathematics camp.

Relationship

One interpretation of developing a relationship with mathematics is grounded in the work of feminists such as Evelyn Fox Keller (1985). Keller wrote of the ways scientist Barbara McClintock maintained relationships with the plants she studied. McClintock felt the need to know a plant deeply from the seedling stage on, including the story of how it grew. Further, she enjoyed this intimate knowing of the plants. In the relationship, there was "engagement and understanding" (Keller, 1985, p. 614).

The engagement aspect of McClintock's relationship with the plants she studied is a concept also found in literacy contexts. Research has shown that engagement is a critical component in making literacy gains; further, the engagement and gains are symbiotically connected to reading genres chosen and the manner in which readers are asked to respond to literature (Deci, Vallerand, Pelletier, & Ryan, 1991; Dole, Duffy, Roehler, & Pearson, 1991; Enciso, 1992, 1996). The basic process of student engagement in learning involves a participant going beyond causal association with an event or inquiry. The process of engaging not only involves acquiring knowledge but also investing oneself in formulating personal questions and connections that will lead to further knowledge development (Morgan & Saxton, 1994; Warner, 1997, 2007).

The understanding aspect of McClintock's notion of relationship is a concept present in the mathematics education community where learning with understanding involves sense-making effort (Baroody & Dowker, 2003; Goos, 2004; Heibert, Carpenter, Fennema, Fuson, Wearne, Murray, Oliver, & Human, 1997). Knowing historical connections, the story, the context; experiencing engagement and understanding; and deeply knowing the content are developed through multiple processes and practices (Common Core State Standards for Mathematics [CCSSM], 2010; National Council of Teachers of Mathematics [NCTM], 2000).

A girls' mathematics camp provides a worthwhile environment in which to consider relationships between program participants and elements of the teaching and learning experience. Three particular forms of relationship, as noted earlier, are relevant to this work: (a) relationship with the content (mathematics); (b) relationship with the pedagogy (the intersection of content, teaching, and learning); and (c) relationship with people (interpersonal relationships among participants and the intrapersonal relationship with the self).

The Matherscize program's research-based design makes the camp curriculum, pedagogy, and nature of participation an appropriate context through which to examine relationships. The NCTM (2000) and CCSSM (2010) provide recommendations for and insights into effective mathematics education in relation to content, processes, and practices. Other mathematics education research offers additional guidance. For example, work involving voice in relation to women and mathematics informs both curriculum and pedagogy (Anderson, 2005; Becker, 1995; Erchick, 2002; Jacobs & Becker, 1997; Karp, Brown, Allen, & Allen, 1998; Kort, 1996). The design of Matherscize is informed by research recommendations that include making activities meaningful, allowing opportunities for development of voice, and maintaining high expectations for student engagement and mathematical understanding (Boaler & Greeno, 2000; Cobb & Yackel, 1996; Martin & Towers, 2009; Nicol & Crespo, 2005; Schoenfeld, 1998).

PROGRAM DESCRIPTION

About Matherscize

Beginning in 1999 and continuing annually, I have conducted Matherscize as a week-long mathematics camp for middle-grades girls, those entering grades 5, 6, or 7 the autumn following the camp (typically ages 10–14). The program is presented as a pleasant experience, as illustrated in the camp logo (see Figure 2.1). In the camp, we teach mathematics through cooperative problem solving and integrated content, with a focus on mathematical processes, namely communication, representation, connections, reasoning and proof, and problem solving (NCTM, 2000). The camp content is intended to augment or enhance–but not replace or remediate–school learning.

Matherscize is community-like in its focus, activities, personnel, and physical facilities. Camp activities provide opportunities for the girls to get to know each other and to support one another's work and growth throughout the week. Interactions with adult participants are casual and most campers socialize with instructors during the lunch period. Free-time

Natalie Wetzel

Figure 2.1 Matherscize program logo.

activities include playing mathematics games and taking bookstore excursions. Camp activities occur each day in the same dedicated space, which becomes a "home" for the program where some daily work is posted and ongoing activities are revisited across the week.

I offer Matherscize free of charge on a first-come, first-served basis and house the program on the Newark campus of The Ohio State University. Through research and programming grants, campus administrative support, and departmental support, I have been able to conduct the camp, pay honorariums for selected experts (in areas such as technology), supply participants with program materials, such as games, books, rulers, and other mathematics tools, and collect and analyze program-related data.

The Matherscize camp starts at 9:00 a.m. and ends at 3:00 p.m. daily for five days. Girls may arrive early or leave late each day up to 30 minutes in order to meet with other campers to play games or to accommodate parent or guardian schedules.

Campers and Volunteers

Matherscize uses multiple approaches to generate enrollment. To help publicize the program, a project web page serves both informational and functional purposes (http://newark.osu.edu/derchick/matherscize). Flyers are also distributed to teachers through various mathematics education programs, campus graduate programs, and email. Finally, campers hear of the camp by word of mouth, often from previous participants, fellow students, and teachers.

The self-selective nature of participation in Matherscize parallels the "community nomination" sampling strategy used by Ladson-Billings (1994) in *The Dreamkeepers*. This strategy "means that researchers rely upon community members...to judge people, places and things within their own settings" (p. 147). In the case of Matherscize, participants trust other members of the "community" of students and teachers to judge the value of the program and its appropriateness as an activity for friends. Trusted community members making recommendations were former Matherscize participants, teachers in our graduate programs, and preservice teachers serving as classroom interns in the schools. Importantly, inherent in the girls' word-of-mouth recruitment of camp participants is an implication that "community nomination" suggests some kind of community validation.

Typically, about 20% of camp participants commute from outside the immediate area, primarily from a large, racially diverse city, and it is not unusual for one or two campers to travel from out of state. However, most campers live in rural and small towns served by the campus. The local community includes minimal racial/ethnic diversity but is quite variable in terms of socioeconomic status. Camp enrollment is representative of the racial, social, and economic composition of the region. Of the 80% of campers who live in the local community, half come from an affluent school district and the other half from poor rural and small urban areas. The girls come to the camp with a wide range of mathematics skills and dispositions. Some love mathematics and want to have more fun with it. Others believe they need help with basic skills or doing better on state-mandated exams.

Adults who participate in Matherscize in some manner are mostly female and come from multiple sources. Some are practicing teachers who engage with the camp voluntarily for professional development. These teachers might teach a lesson, work with small groups of campers on activities and projects, support curriculum revision, generate program materials, or supervise noninstructional time. Graduate student volunteers might also seek professional development experiences and perform some of the same duties as teacher participants, but they might gain research experience as well. Parent and community volunteers work with or supervise campers, generate and organize materials, and support the camp's technology needs. Matherscize also incorporates past campers as volunteers, some for multiple years into and through high school. These volunteers manage digital materials, help support campers, and design and teach selected lessons. Returning campers sometimes discuss their college work in mathematics-related fields and, in doing so, inspire campers. All volunteers receive training prior to the camp, but they come to us with a vested interest in, belief in, or prior experience with Matherscize. The camper-to-instructor ratio is about 7:1, but the camper-to-staff ratio that includes other adult participants (e.g., volunteers) is about 3:1.

Camp Activity

One distinctive feature of Matherscize is the intensive nature of the campers' experience. The camp has a low camper-to-adult ratio, as noted above. The camp enrollment is intentionally small, usually between 25 and 40 girls, allowing the camp to provide individualized attention for participants to maximize learning and to integrate into the camp community. The curriculum is rigorous and ambitious (one parent noted that her daughter "did more math in a week than she did all year at school"), but the schedule allows plenty of time and support for project completion and deep reflective discussion. These factors collectively create the intensive Matherscize experience.

We teach the girls the following games during the camp: Mancala (logic), 24 Game (number skills), SET (visual perception), and Old Bachelor (women's career possibilities). The girls choose from these games to play during the lunch hour or before or after camp each day.

The campers and volunteers either pack a lunch or purchase it on campus Monday through Thursday of the camp week. Campers choose where they would like to eat their lunch—the picnic tables immediately outside the classroom, the campus cafeteria, a specific social area in the education building, or a large covered pavilion on campus. Volunteers accompany the girls to supervise them and eat lunch with them. On Friday the program provides lunch, and campers and volunteers often choose to contribute fruit, desserts, or snacks to the menu.

The camp's pedagogical approach is student-centered and grounded in problem solving. We thus design our closing activities to align with this approach. Instead of the campers presenting their work at camp closing as a kind of show-and-tell, campers staff centers where they teach guests (parents, siblings, extended family, childcare providers, and neighbors) what they learned in the way they learned it. Each camp activity has its own center, and guests rotate among the centers to experience all of the activities in abbreviated form.

Program Approach to the Content Component

Program content is grounded in mathematics education research on teaching, learning, and educational equity. For example, for one camp activity described below, Matherscize uses NCTM (2000) and CCSSM (2010) expectations that students be able to explore mathematics in context and to analyze and represent data, and in its equity principle, the NCTM promotes a vision of "high expectations and worthwhile opportunities for all" (p. 2). The example involves a Matherscize lesson using *Swamp Angel* (Isaacs, 1994), adapted from "Feisty Females: Using Children's Literature with Strong Female Characters" (Karp, Allen, Allen, & Brown, 1998). This lesson includes opportunities for learning about the mathematics and reflecting upon social concepts relevant to girls' development.

Swamp Angel is a tall tale where the main character, Angelica, is a 200-foot-tall young girl who lives in Kentucky. The camp activity employs this children's literature as an entry point for critique, analysis, and data manipulation. Spreadsheet data are adapted from that provided in the "Feisty Females" article and are presented to Matherscize campers along with data inserted from *Swamp Angel.* The inserted data consist merely of the addition of a line for Angelica that included the entry of 200 under height but left all of the other cells across the row blank for the campers to complete. The original data are students' body measurements, recorded in a spreadsheet without units. The data from the article are inconsistent in that, as reported in the "Feisty Females" article, some children in their work measured in the metric system and some did not, or some erred by recording a height of 4 feet 8 inches as 4.8. Because of irregularities in the data set, the addition of Angelica's data does not disrupt the set in noticeable ways. In Matherscize, the girls are expected to explain the irregularities in the data set and to complete missing data in the spreadsheet.

Full class discussion of the story precedes the data work. The girls bring to that discussion their thoughts about how it feels to be different like Angelica, to what degree it is important to be liked and have friends, and how Angelica won respect with her strength of character and body. Thus, "Feisty Females" provides a way to use children's literature with strong female characters to support gender equity, while *Swamp Angel* specifically provides a context for working with data and discussing cultural and societal expectations for girls' appearance and behavior.

Program Approach to the Pedagogical Component

The Matherscize pedagogy is process-based in that process is the main focus for engaging content. This focus impacts instructional decisions and lesson implementation in multiple ways. First, mathematics topics and concepts from both the NCTM (2000) and the CCSSM (2010) and the NCTM's mathematical processes form the foundation for all lessons. The girls complete all tasks in cooperative groups, where communication, problem solving, reasoning, argument, logic, connections, and multiple representations are processes used for making sense of the mathematics. Individual lessons are designed explicitly to approach the content by way of selected processes. For instance, in a geometric transformation lesson, quilting was a context, or real-world connection, in which the girls designed quilts while learning about translation, reflection, and rotation. In another lesson, campers connected the processes they employed in a collaborative writing task to the processes they use to jointly solve mathematics problems. Groups wrote stories together based on *The Mysteries of Harris Burdick* (Van Alsburg, 1984) and generated a list of the processes they had used. They then revisited the list to determine which processes also applied to solving a mathematics problem collaboratively. Finally, they solved some mathematics problems in groups and then returned to the list to verify that indeed most or all of the processes (sometimes with slight language changes) applied to both writing and mathematics.

Program Approach to the Inter/Intrapersonal Component

The interpersonal component of Matherscize involves the girls relating to other camp participants (other girls and adults) and the intrapersonal includes the girls relating to themselves. In Matherscize:

1. Collaborative working groups provide opportunities for the girls to work together, get to know each other, and help each other. The girls are assigned to groups that remain the same during the week, and an initial discussion during the camp makes it clear that the Matherscize team expects everyone to care for and about each other, especially within the working groups. We talk with them about a kind of buddy system, where they each take responsibility for helping girls in their group. If someone does not understand directions, group members help or call on a staff volunteer to assist; if someone does not feel well, group members seek help; if someone in the group is dominating or not participating, group members take responsibility for helping that girl participate appropriately and, if needed, call on a volunteer for assistance. Groups are carefully formed to avoid having girls who are in the same school and grade in the same group so that they have to get to know everyone in their group as a new "friend." Although we group by age and grade, placing together girls who are close in age often results in group members spanning two grade levels.

2. The presence of numerous volunteers and the lesson design provide a context for the girls to share their work and reflect on and reconsider solutions and approaches in a low-risk environment. Because of this "safe" setting and the extended time given to activities, the campers tend to give deeper explanations in small-group and whole-class discussions. They openly generate group corrections and clarifications.

3. Instructional approaches such as think-pair-share involve situations where the girls first reflect on material individually (think), then discuss their thoughts with another girl (pair), and finally share their ideas with the working group (share). (A modification is also used where "pair" is done as a small group.) This strategy of moving from autonomous thought to paired discussion to group sharing provides a graduated opportunity to surpass uncertainty and encourages the girls to learn to listen to themselves and to each other. Joint participant activities are intended to help each girl experience doing mathematics in a community of other girls, a community to which they can feel they belong. Private reflection is designed to encourage girls' thoughtful consideration of their relationships with mathematics and of important program emphases.

The "Data Faces" lesson is an example of a community-building activity. In this lesson, the girls design a survey instrument to learn about each other. The instructor facilitates group development of a survey tool, starting with generating possible questions campers might have about their new peers in the camp. During this part of the lesson, the girls learn some information about the rest of the campers while making decisions about what is most important to know. Survey questions typically center on the girls' favorite hobbies, books, music, or food, what kind of pets the campers have, or how many siblings each girl has. The teacher facilitates the girls' streamlining the list to five questions, and campers design an icon for each possible answer to the survey questions. Once the survey tool is designed, the campers individually respond to the survey by choosing the icon that represents their response and drawing it on a face in predetermined locations. This results in a "Data Face" for each girl from which they can read raw data from all participants' data faces and make graphic representations to display collated data. In addition to providing an opportunity to work with real data and multiple representations of those data, this lesson helps the girls learn about each other, reflect upon their own interests, and choose what they want to share with the other participants. Other camp activities also encourage intrapersonal reflection through synthesized group presentations and individual writing prompts.

PROGRAM EFFECTIVENESS

Methods

I conducted this inquiry from the interpretivist perspective in both data collection and analysis (Denzin, 1989; Schwandt, 1994). What that means for this work is that even though prior research in the field informed analysis, I also heavily weighed perspectives and interpretations of all camp participants. Thus, the story of the girls' experience is grounded in the perspectives and interpretations of the participants: 32 girls, 20 parents, and 7 teachers. Three graduate research assistant volunteers contributed data in the form of research reflections.

Earlier in this chapter I noted the sampling strategy for participation in the Matherscize camp. Ladson-Billings' "community nomination" provided a kind of community validation in participant self-selection. For participation in this study, no additional sampling strategy was used because the entire population was invited to participate. All camp teachers and research assistants were invited to participate via an institutional review board (IRB)-approved consent form where they could choose to participate or opt out of the research without their decisions impacting their participation in the camp. All parents—for themselves and their daughters—approved their participation in the research with an IRB-approved consent form prior to the start of the camp. Additionally, all campers were asked orally during the camp if they would be willing to participate. All campers, teachers, and research assistants consented to take part in the study. Additionally, 20 parents (or guardians) from the 32 campers chose to contribute responses to the parent feedback form, the source of parent data for the study.

Data collected from responses to a parent feedback form were collected at the closing camp session or returned by mail within the next week. Given that all girls had at least one representative in attendance at the closing, we were assured that all parents had the opportunity to complete the form. As noted, 20 parents chose to participate. Questions on the parent feedback form included requests about how the parents learned about the daily camp activities (e.g., from the camper, from the closing session, or from another source). Parents were also asked open-ended questions about their sense of the importance and value of the camp activities for their daughter's mathematics growth, what impressed them about the camp experiences, what else they would have liked their child to experience, and any other comments they might want to add.

All data collected for analysis were typically gathered during camp implementation and evaluation, including girls' written reflections, work products, and parent and teacher feedback. Table 2.1 outlines the full set of data collection activities.

TABLE 2.1	Matherscize Data Collection Activities
Student	Initial application essay; daily written responses on camp activities; products resulting from camp activity: Data Faces graphs; Harris Burdick writing/ problem-solving process discussion; fraction collaborative problem solving, whole-group solution sharing; Swamp Angel data analysis, representation, presentation; quilting projects; simple machine projects; aerodynamics projects; informal games discussions; feedback at the end of the camp activities
Teacher	Daily written reflections; daily teacher debriefing sessions; follow-up feedback at the end of camp activities
Parents	Parent feedback upon completion of the summer camp

Data Analysis

One research assistant and I performed all data analysis, first synthesizing all data and then independently reading and coding all data as pertaining to (M)athematics, (P)edagogy, or (I)nter/Intrapersonal topics. These narrative data were identified as data points by topic. Any segment of narrative could contain multiple data points, and any data point could be assigned multiple codes. For example, a teacher wrote of the value of the small-group collaboration in terms of helping the girls learn more about each other, and then wrote of how that collaboration fell apart when the activity focused more on the mathematics. This data sample has two data points: one about the value of collaboration (coded P and I) and the other about how it deteriorated when mathematics came into play (coded I and M). Data points were then sorted and arranged by code.

Reliability

This inquiry is grounded in a program that I developed and implemented. That fact introduces risk in terms of this study's reliability, the definition of which I take from Denzin and Lincoln (1994) as "the extent to which findings can be replicated, or reproduced, by another inquirer" (p. 100). In an effort to limit the extent to which this inquiry might become a victory narrative (Cary, 1999) of Matherscize, I used several strategies to achieve more reliable results. First, I hired a graduate research assistant, not connected to my work in any other way, to oversee data collection. An undergraduate student assisted with this task. The graduate research assistant also assisted in organizing and synthesizing the data and then independently (but simultaneous with me) coding the data. We then collaboratively reviewed our initial findings, and through mutual analysis and discussion, settled differences. Second, I solicited peer review of the synthesis, analysis, and write-up of the study.

Findings

Relationships With Content

The majority of the Matherscize campers, in their application essays, belie the perception that girls do not enjoy mathematics. Certainly, the participants self-selected, but some did so after being urged by family or teachers. Still, all of the 32 girls who attended the camp during the period of this study either indicated that they had positive relationships with mathematics or that they wanted to. All but 5 wrote in their application essay that mathematics was a favorite subject, "exciting," or otherwise enjoyable, specifically mentioning geometry, algebra, or problem solving. One wrote, "When I do problem solving it makes me relax and stretch my knowledge in math." The girls wrote of enjoying puzzles and brainteasers or of having fathers create word problems for them "just for fun." One girl wanted to become a mathematics teacher.

Of the 5 girls who did not write of positive relationships with mathematics, it was not that they did not want to have one. Four either explicitly claimed that mathematics was hard for them or wrote that they needed help to improve their grades. Of those, one indicated that she had "yet to earn an A" and another who reported a grade of B+ in math said she "need[s] help in math." Another who did not write of a positive relationship with mathematics wrote, "This year is a struggle in math because now my teacher is putting letters in problems and it gets very complicated . . . I would like to attend math camp because I think it will help me see math differently."

Following the camp experience, the girls reported that mathematics was learned in a fun way in the camp, which pleased them. They explained that the camp activities "taught us a lot of math without us knowing it" and "It doesn't seem like it's about math! I really liked it." The campers developed a more enjoyable connection to mathematics, which they considered a good thing. Thinking it was fun and having their perception changed in that regard was clearly important to them, but data from volunteer reports revealed a more serious side to the campers' changed perspective. Yes, it was fun for the girls, but camp volunteers who assisted the girls in preparing the teaching centers for the closing activity reported that every camper experienced a kind of "aha" moment about mathematics as they prepared to teach their guests, which was that they engaged with serious mathematics in their week of camp. Campers came to realize that developing a survey tool and working with raw data was "real math" and that improving logic while playing Mancala was important and challenging mathematics. They expressed to some volunteers that thinking about problem-solving processes made them better "mathematicians" and that working together made them think more about mathematics.

The description of the girls' relationship with mathematics is also informed by parent and teacher feedback. That feedback tells us that for the parents and teachers, a positive relationship includes engaging in challenging interactions and numerous and varied mathematical activities. It also involves developing better dispositions, such as recognition of the role of mathematics in one's life and the value of learning mathematics. I explain these elements in the paragraphs that follow.

Parent and teacher responses indicate that a necessary element for developing a strong, positive relationship with mathematics is providing program tasks that are challenging. Every parent comment noted the importance of the mathematics being challenging. Although one parent thought the activities were not challenging enough, all other parents—as well as teachers who provided feedback—considered the activities challenging while also being fun and relevant. A characteristic parent response was: "I think it encourages and challenges her when she could easily decide to dismiss math since it comes easily and she isn't always challenged." Similarly, multiple teacher comments related how actively involved the girls were in challenging activities.

In addition to the activities being engaging and relevant, the act of doing mathematics is itself a necessary part of developing a relationship with the content. As one's relationship with mathematics develops through challenging interactions with content that has a valuable role in life, engaging in many and varied mathematics activities enhances the relationship. From the camp experience, parents and teachers appreciated the girls' multiple opportunities to interact with and relate to mathematics. Parents "knew [the girls] were learning important scientific and mathematical principles" and left the camp "with a great impression of the Matherscize activities . . . [They did] in one week what should have been done in a whole school year." One parent liked "all of the different activities they got to do" and another that the camp offered a "continuity of concepts . . . learned in school" that would help the girls "remember in a fun way."

The girls' relationship with mathematics was not simply about products and concepts. It was also about dispositions. Further support for what it means to have a relationship with mathematics comes from parent descriptions of the role they believe mathematics plays in their daughters' lives. Parents commented that the camp helped their daughters understand the value of mathematics, giving them insight into:

- "the importance of math and its usefulness,"
- "how math can be used in their lives,"
- "an interest in math and understanding its practical relevance to her life," and
- how "math is applied well beyond where she would actually expect it."

Parents and teachers also recognized selected dispositions, in particular, confidence and enthusiasm, that they found to be a valuable consequence of the girls' mathematics work during the camp. These dispositions can strengthen the girls' relationship with mathematics and their own intrapersonal relationships (the latter of which I address later in this chapter). One parent wrote about it in terms of a "confidence that my girls seemed to have in themselves every night and the excitement they showed when they told me about their day." Another parent commented that her or his daughter had never had so much fun with math and wrote, "I enjoyed her explaining what happened every day. But seeing [in the closing presentations] was awesome." Parents and teachers enjoyed the enthusiasm they saw in the girls and explained with comments such as, "She was enthusiastic about what she had done (at camp and for hours after camp—not just the initial sharing after pick up from camp)," "What most impressed me was that my daughter wanted to go back each day, and must have worked very hard because she took afternoon naps on several days," and "It was so neat to see how much confidence the girls had by the end of the week."

Relationships With Pedagogy

As indicated earlier in this chapter, process-based pedagogy grounds the camp's lesson design and teaching. The girls' response to this approach sometimes indicated acknowledgment and appreciation, but at other times resistance. In their application essays, the girls showed no evidence of anticipating a process-based approach to the camp activities. The girls were beginning to know mathematics as something more than computation, as is evidenced in their comments about algebra, geometry, and "numbers and pictures." Still, the girls saw mathematics as skills-based content, mostly focused on doing mathematics in a traditional sense by using algorithms to solve problems and performing the operations of addition, subtraction, multiplication, and division. On the camper final feedback forms, some wrote that a strength of the camp experience was that the girls could "do things the way they want." They could "be creative" and do "problem solving in fun ways."

Participants reported their perspectives on selected instructional processes. Teachers recognized the "valuable" role of communication during the week's activities. Similarly, parents and girls tended to appreciate opportunities to debate and discuss work in groups. However, explicitly connecting the role of processes to mathematics learning resulted in some camper resistance. For the activity described earlier, the girls completed a collaborative writing task, defined the processes they employed, and then applied those processes to collaborative mathematical problem solving. Teachers commented that the girls readily participated in the writing activity and enjoyed reflecting on the processes they used. However, when the activity switched to mathematical problem solving, many girls resisted or rejected the processes the girls had so freely participated in for writing and were to be used again for mathematical problem solving. Because individuals were not interviewed regarding their perceptions of these events, it is not clear how many girls resisted the process focus during the mathematics component of the lesson, or why they did so. Still, every group of four girls struggled with the collaborative focus of the problem solving. It became evident in the follow-up class discussion that most girls either switched to a solitary mode, solving problems in isolation and avoiding communication, or they avoided the problem completely, waiting for peers to complete the task.

Parents appreciated the conceptual connections that formed one element of the process-based pedagogy. They also recognized the importance of the real-world connections and opportunities for creativity that were built into the camp activities. A representative parent comment is, "I think the things my daughter worked on made her think in many different ways, important for problem solving and aiding creativity." Parents were happy to see their daughters "being able to solve problems and seeing that there may be more than one way to achieve something." One parent noted: "She is very creative but not as 'quick' at math as her brothers. I believe giving her the exposure and experience of looking at math concepts and relationships and principles in a creative way is beneficial."

One parent reported uncertainty about the purpose of the writing-process/problem-solving lesson, and another indicated strong disappointment in the lack of focus on computation (the same parent who felt the program was not sufficiently challenging). However, all remaining parents who chose to offer general comments expressed satisfaction with the instructional approaches, particularly the use of methods that went beyond traditional mathematics teaching while still connecting to school mathematics.

Inter/Intrapersonal Relationships

The importance of interpersonal relationships to the girls in this camp became clear in the girls' written responses after the *Swamp Angel* activity. As explained earlier, this activity used a particular piece of literature with a strong female character as an entry point for data analysis, representation, and interpretation. The overall context of the activity included discussion of "tall tales" and respect for other cultures. The heroine in the story is a 200-foot-tall, young Kentucky woods-woman, Angelica Longrider. She earned the name "Swamp Angel" from one of her many kind acts and was a big, strong, talented, and good-hearted member of the community. She withstood some ridicule from selected members of the community, not because of her size, but because of her gender. Girls should be home quilting and making pies, the story's community members said, not out fighting bears, as Angelica did.

In their responses to writing prompts at the end of the day on which they participated in the *Swamp Angel* activities, two-thirds of the girls wrote that they would not want to be like the heroine Angelica because she was "too big," would be "fat," "would scare people," was "too rough," and was "not like 'normal' girls." These campers wrote that they would not want to be the center of attention or be left out or not have things like shoes, clothes, and toys (because they would be too big to find these items in an appropriate size). For the one-third of the girls who did want to be like Angelica, they were concerned with being liked and wanted to have friends and "be respected by all." It is thus apparent that middle-grades girls in general are concerned with fitting in with peers.

The girls anticipated a fun camp experience and indeed reported it to be such at camp end. Some explicitly associated fun with learning about and doing mathematics. However, in other statements it was not clear that the girls were referring to the mathematics when they talked of the camp being fun. Statements such as "I would like to attend because it sounds very fun" could have related to a girl's interactions with mathematics, but they might have instead been based on interactions with other girls, doing engaging activities, spending time on a college campus, or other such possibilities. In this ambiguity about what was meant by "fun," more than one-third of the girls expressed in their application essays an expectation that the camp would be "fun," "exciting," a "great experience," "enjoyable," or "neat." One girl who had attended Matherscize the previous year stated, "I came to Math Camp last year and had a blast! I had so much fun, I invited my cousin to come." Camper feedback at the end of the camp was nearly all positive, although still often ambiguous. The girls wrote that they enjoyed camp, wanted to return next year, "liked everything about the camp," and that "it was great fun." One girl wrote, "I would like to see this program for older kids since this was my last year."

Despite some ambiguity in the girls' written comments, some statements in the application essays clearly connected fun to interpersonal relationships within the camp, in particular, peer relationships. Some reasons expressed for wanting to attend the camp included that the girls wanted "to meet other girls with the same interests," that it "sounds fun to learn and hang around all girls for a whole week," and that they hoped they would "make new friends" and "learn a lot while making new friends." The girls looked forward to "doing many different group activities" and being in the camp "because some of my friends might be there." Some girls particularly wanted to attend because the camp was for girls only. These girls made comments such as "I would love to go to the Matherscize camp to meet new friends and there will be no boys" and "I wanted to attend an all girls camp because boys play to [*sic*] much in class and it is hard to concentrate."

Following the camp, the girls, parents, and teachers noted the importance of interpersonal relationships formed during the camp. Although some camper groups had difficulty working together (illustrated by one girl's comment "Don't listen to Betsy"), all participants listed group work (e.g., "team work" and "working with other girls in a group solving problems") as a program strength. Some girls explained that the program "helped me meet new people and have fun conversations" and that they "really liked the whole experience. It was a fun way to learn." Teachers commented on "great group work, verbal communication about math" and that "many of the girls . . . really bonded and made new friends—they were signing each others' journals."

Parents cited the girls' interpersonal relationships with the staff as another strength of the camp experience. The girls found the teachers easy to communicate with, making statements such as, "The teachers are gentle and nice . . . even when they woke up on the wrong side of the bed." One parent reported "interaction with the staff" as a favorable aspect of the program, particularly in light of the low camper-to-teacher ratio.

Across participants, the fact that the camp was for "Girls, only!" (parent comment) was another strength. In addition to comments such as those noted earlier regarding some girls' favorable perceptions of a program with no boys and especially none to distract them, the girls offered further commentary about the perceived value of an all-girl environment. They stated that "it was a fun experience to just work with girls" and that "it was just girls so some of them understand you and then you might learn things you might not have known." The girls thus indicated an appreciation of a girls-only learning opportunity. Teachers also commented on the all-female program, with one teacher listing one program strength as having "fun, hands-on, 'minds-on' activities, small groups, plenty of helpers particularly in an all-girl environment."

In addition to interpersonal relationships among campers, participants identified two other aspects of inter/intrapersonal relationships as program strengths. One was the overall environment, where a sense of community provided a safe environment for learning. One parent explained in relation to her daughter, "She has always excelled in this area [of mathematics]—It is wonderful she has a place to explore and have fun." Another appreciated that "this camp was a great experience for the girls to work cooperatively in a risk-free environment." Teachers recognized that the supportive community atmosphere made presentations of group work during the week an activity with little emotional risk in terms of possible errors or failure, and the pedagogical choice not to have the girls present to the whole group for certain activities allowed for further safety when working with challenging concepts.

   The other type of personal relationship identified in participant writing is internal relationships with the self (intrapersonal). In the earlier discussion about the girls' relationship to the content, I described strengthened dispositions toward mathematics. The confidence, excitement, enjoyment, enthusiasm, and engagement identified by parents and teachers do not simply stem from the girls' developing relationship with mathematics. They are also outgrowths of the girls' developing relationships with themselves. These girls, despite often entering the camp with a positive perspective toward mathematics, show what parents identified as stronger characteristics of their daughters at program end. Two sample parent comments reported earlier but which apply here are "confidence that my girls seemed to have in themselves every night and the excitement they showed when they told me about their day" and daughters who "were excited to go to each day. That means a lot to me." The girls' relationship with themselves—their confidence while in the mathematics environment—was apparent.

DISCUSSION

The girls who participated in this camp were not a random sample, but neither was participant solicitation purposeful. Because of their choice to attend Matherscize, the girls comprised a special group, one that is self-selected and belies the myth that girls cannot do, do not do, or do not want to do mathematics. Despite economic and skill differences across the group, the girls shared a desire to interact with—or at least improve their relationship with—mathematics.

   All that the participants valued—the problem solving, the opportunity to be creative, communication, group work, meaningful applications—occurred in a special context. The environment was identified as low-risk and was favorably acknowledged for being an all-girl setting where individual differences were respected. The girls liked the instructional approaches and enjoyed the social aspect of interacting with other campers and playing games during lunch and in the early morning in what became a community in only a few days.

   Working through the program content, instructional approaches, and social expectations in the camp community presented a few problems. The girls were placed in groups of four (randomly within grade level) and remained with their group for the duration of the camp. One program objective was to support the girls in developing all of their relationships during the camp—inter/intrapersonal and with content and pedagogy. These relationships often converged in the small-group settings. Two groups in particular had difficulty in their interactions. This occurred when multiple strong personalities within a group each considered their own problem-solving method superior and created barriers for the group. The program vision was that the girls—with some staff guidance, including facilitated reflection—would make the group dynamics more functional. Unfortunately, the short duration of the camp became prohibitive for resolving such dynamics.

The teachers noted the issues some groups experienced in working together. They were troubled by the low quality of the project products for the two groups who had difficulty working collaboratively. They were also concerned about the amount of time the girls in these two groups "wasted" in interacting and debating. They suggested that either the teacher can: (a) "Possibly give the girls some strategies on how to work together as a group and what to do when a problem arises;" or (b) "perhaps change groups—mix it up?" Interestingly, if one goal is to have the girls learn how to work with their group and participate in the overall community, the teachers' solutions do not particularly fit program intent. The first suggestion might be appropriate if the girls were given an opportunity to construct a solution, one facilitated by the teacher rather than one "given" to the girls. The second solution, in which the teacher would change group membership, is more of a management decision than a pedagogical one. Such a solution would result in a more smoothly running classroom, but the girl(s) in question would not have learned how to negotiate a relationship with their peers, and the opportunity to group these girls to serve a pedagogical purpose would have been lost.

The teachers met with the camp leadership at the end of each camp day and discussed issues. It was at this time that we as a group decided to help the girls learn some strategies for working together and to avoid changing the makeup of the groups. Because it was a goal of the camp for the girls to learn to work collaboratively, that option seemed most appropriate. Teacher feedback at the close of the camp, however, remained conflicted as noted, with some preferring to change the groups rather than taking time to teach the girls skills for working together. In subsequent camps, we included in our teacher training how to help the campers develop skills to work together and pressed the point that changing group makeup was not an option. Those decisions resulted in no significant collaboration problems in groups for later camps.

In addition to social processes, Matherscize camp lessons maintained a strong focus on the mathematical processes and on the girls' recognition of those processes. Some potential conflicts arose within that focus. Teachers reported they would have liked to have spent less time on the "lead in" and more time on the mathematics, even though (a) parents and teachers alike were impressed with the amount of mathematics addressed in the camp and (b) the girls liked the process component of the lessons, which kept them engaged and active. The only time the girls resisted or disengaged was when we moved the emphasis from process to content.

This brings me to the issue that mathematics still does not adequately fit into the lives of many women. In Matherscize, the young girls and their parents appreciated the contextualized mathematics presented in the camp, particularly that the content was made relevant through real-world connections. This begs the question of what happens between the middle grades and adulthood, or simply outside of a positive experience such as this, that moves girls away from being comfortable and excited about mathematics to where they do not connect with it.

Social aspects and instructional processes aside, the girls' interactions with the mathematics content presented some cause for thought. For one thing, the girls stayed engaged in process-oriented lesson components. They were less engaged in parts where new learning, application, or other direct work with mathematics was required. In these activities, their analyses often lacked depth, and the groups tended to hurry through task development and presentation. In the lesson on geometric transformations in a quilting context, the girls understood the transformations and were able to recognize and apply specific transformations. However, when creating their own quilt design, the girls resisted application of the transformations. The final product required that the girls design a square to be transformed each of three different ways in a predetermined order. Nearly every girl challenged teachers in the requirement to use all three of the transformations and to do so in a specified order. Two girls explicitly chose not to follow the directions. As was clear in the conversations around the quilt design, these two girls, as well as those who followed the directions in protest, prioritized their own sense of aesthetics over the mathematics. They wanted their four-square quilt to be symmetrical/"normal," but following the directions led to a quilt that was not symmetrical. These girls neither wanted to follow the mathematics nor were interested in the reasons given for using the mathematics and developing an asymmetrical quilt. We knew from experience that it was in the extended activity, where campers designed large digital quilts with images of their individual four-square quilts, that the symmetrical quilt yielded a less interesting digital quilt. In the following years, we introduced examples of the final digital quilt early in the lesson and encountered less resistance for following the mathematics requirements and accepting an asymmetrical four-square quilt.

The teachers in this program recognized that the girls readily participated in the writing/process lesson but resisted a process focus when they arrived at an "equal-shares" mathematics problem that involved use of fractions. Although the problems were solvable with computation skills the girls should already have possessed, the mathematics tended to be challenging. Additionally, the task required that the girls use models or drawings to explain their solutions conceptually. This, too, challenged the girls. Even though no girl identified these problems as either challenging or easy, it was here that the majority disengaged with the lesson.

Nearly all of the girls demonstrated skill in procedural solutions, but their conceptual solutions were limited in terms of variety and creativity. They often wanted to simply get the mathematics done. Doing it to the best of their ability, or finding creative or innovative solutions, did not interest many girls.

Some potential for further study rests in investigation of the girls' perceptions of the nature of mathematics, which might explain some resistance to the mathematics in Matherscize. Girls who associated mathematics with computation and "not letters or pictures" encountered something different in Matherscize. Those who wrote about liking Matherscize mathematics with comments that this "doesn't seem like it's about math! I really liked it" makes one wonder how that perception impacts future learning, especially in more traditional settings.

CLOSING COMMENTS

We do not fully understand the complex context of the intersection of gender and mathematics, but we do know that mathematics is still, in spite of progress made, not experienced as a particularly appropriate part of many women's lives. In Matherscize, however, the effort to support the girls' developing relationships with mathematics was intended to be holistic. Where research in mathematics education and on gender and mathematics provides insights into elements necessary for successful mathematics development, it is unclear how those elements work together. In Matherscize, the mathematics was experienced as a relevant part of the girls' lives. The environment of the camp—the community and the relationships developed within it—are integral to that relevance. The girls claimed they had fun in the camp, but they came to the camp expecting and wanting to have fun. The girls brought to their experiences in the camp an interest in improving their relationship with mathematics, and they indicated growth in their appreciation, engagement, interest, and excitement about the work they did in the camp. The camp community allowed that to happen, and although further inquiry into content learning in a context like this would be appropriate, scholarship already suggests the benefits of learning in such a community (Bryant, 1999; Goos, 2004; Vygotsky, 1978; Wood, Cobb, & Yackel, 1992).

Learning mathematics in context suggests further inquiry. The girls in this program enjoyed the mathematics contexts in the lessons and remained engaged as they progressed through the process components. Thus, the purposes and outcomes of the camp—the developed community, the growth of inter/intrapersonal relationships, the process-based lessons, and the relevant contexts, all supporting development of varied types of re-lationship—successfully brought the girls to important points of encounter with the content. However, it was at those points that resistance appeared within the relationship between mathematics and these middle-grades girls. In these cases, unknown elements, perhaps the mathematics itself, created barriers that resulted in resistance to relating to and intimately knowing the content. Given that the processes and community helped these girls inter-act with the content, the question becomes: Why did the girls stop at the mathematics? Thus, in addition to future inquiry into mathematics learn-ing that transpires in a community such as Matherscize, inquiry into better understanding the elements of relationship at points of resistance would likely help mathematics educators better support student negotiation of those elements in terms of how to surpass them and connect to and relate to the mathematics content.

Beyond the realm of this study, but an emergent issue, is the task at hand for teacher education. Teacher responses about a process-oriented "lead in" being wasted time contradicts their additional observations that the processes helped the girls communicate and complete their work. Teacher education will need to take up the task of helping teachers reconsider a stance that process is a "lead-in" activity instead of a valuable element that supports learning.

REFERENCES

Anderson, D. L. (2005). A portrait of a feminist mathematics classroom: What ado-lescent girls say about mathematics, themselves, and their experiences in a "unique" learning environment. *Feminist Teacher, 15*(3), 175–194.

Baroody, A., & Dowker A. (Eds.). (2003). *The development of arithmetic concepts and skills: Constructing adaptive expertise.* Mahwah, NJ: Erlbaum.

Becker, J. R. (1995). Women's ways of knowing mathematics. In P. Rogers & G. Kai-ser (Eds.), *Equity in mathematics education: Influences of feminism and culture* (pp. 163–174). London, England: The Falmer Press.

Becker, J. R. (2002, April). Personal communication.

Boaler, J., & Greeno, J. G. (2000). Identity, agency, and knowing in mathematics worlds. In J. Boaler (Ed.), *Multiple perspectives on mathematics teaching and learn-ing* (pp. 171–200). Westport, CT: Ablex.

Bryant, C. (1999). Build a sense of community among students with student-cen-tered activities. *The Social Studies, 90*(3), 110–113.

Cary, L. (1999). Unexpected stories: Life history and the limits of representation. *Qualitative Inquiry, 5*(3), 411–427.

Ceci, S. J., & Williams, W. M. (2010). Sex differences in math-intensive fields. *Current Directions in Psychological Science, 19*(5), 275–279.

Ceci, S. J., Ginther, D. K., Kahn, S., & Williams, W. M. (2015). Women in science: The path to progress. *Scientific American Mind, 26,* 62–69.

Cheryan, S. (2012). Understanding the paradox in math-related fields: Why do some gender gaps remain while others do not? *Sex Roles, 66*(3), 184–190.

Cobb, P., & Yackel, E. (1996). Constructivist, emergent, and sociocultural perspectives in the context of development research. *Educational Psychologist, 31*(3/4), 175–190.

Deci, E. L., Vallerand, R. J., Pelletier, L. G., & Ryan, R. M. (1991). Motivation and education: The self-determination perspective. *Educational Psychologist, 26*(3), 325–346.

Denzin, N. K. (1989). *Interpretive interactionism.* Newbury Park, CA: Sage.

Denzin, N. K., & Lincoln, S. (1994). *Handbook of qualitative research.* Thousand Oaks, CA: Sage.

Diekman, A. B., Brown, E. R., Johnson, A. M., & Clark, E. K. (2010). Seeking congruity between goals and roles: A new look at why women opt out of science, technology, engineering, and mathematics careers. *Psychological Science, 21*(8), 2051–2057.

Dole, J. A., Duffy, G. G., Roehler, L. E., & Pearson, P. D. (1991). Moving from the old to the new: Research on reading comprehension instruction. *Review of Educational Research, 61*(2), 239–264.

Enciso, P. E. (1992). Creating the story world: A case study of a young reader's engagement strategies and stances. In J. Many & C. Cox (Eds.), *Reader stance and literary understanding: Exploring the theories, research and practice* (pp. 75–102). Norwood, NJ: Ablex.

Enciso, P. E. (1996). Why engagement in reading matters to Molly. *Reading and Writing Quarterly: Overcoming Learning Difficulties, 12*(2), 171–194.

Erchick, D. B. (1996). *Women and mathematics: Negotiating the space/barrier* (Unpublished doctoral dissertation). The Ohio State University, Columbus.

Erchick, D. B. (2002). "The Square Thing" as a context for understanding, reasoning and ways of knowing mathematics. *School Science and Mathematics, 102*(1), 25–32.

Good, C., Rattan, A., & Dweck, C. S. (2012). Why do women opt out? Sense of belonging and women's representation in mathematics. *Journal of Personality and Social Psychology, 102*(4), 700–717.

Goos, M. (2004). Learning mathematics in a classroom community of inquiry. *Journal for Research in Mathematics Education, 25*(4), 258–291.

Heibert, J., Carpenter, T., Fennema, E., Fuson, K., Wearne, D., Murray, H., Oliver, A., & Human, P. (1997). *Making sense: Teaching and learning mathematics with understanding.* Portsmouth, NH: Heinemann.

Herzig, A. H. (2002). Where have all the students gone? Participation of doctoral students in authentic mathematical activity as a necessary condition for persistence toward the Ph.D. *Educational Studies in Mathematics, 50*(2), 177–212.

Issacs, A. (1994). *Swamp angel.* New York, NY: Sutton Children's Books.

Jacobs, J. E., & Becker, J. R. (1997). Creating a gender-equitable multicultural classroom using feminist pedagogy. In J. Trentacosta & M. J. Kenney (Eds.), *Multicultural and gender equity in the mathematics classroom, 1997 Yearbook* (pp. 107–114). Reston, VA: National Council of Teachers of Mathematics.

Karp, K., Allen, C., Allen, L. G., & Brown, E. T. (1998). Feisty females: Using children's literature with strong female characters. *Teaching Children Mathematics, 5*(2), 88–94.

Karp, K., Brown, E. T., Allen, L., & Allen, C. (1998). *Feisty females: Inspiring girls to think mathematically*. Portsmouth, NH: Heinemann.

Keller, E. F. (1985). *Reflections on gender and science*. New Haven, CT: Yale University Press.

Kort, E. (1996). Examining the horizons of young women with worthwhile mathematical tasks. *Focus on Learning Problems in Mathematics, 18*(1–3), 138–145.

Ladson-Billings, G. (1994). *The dreamkeepers: Successful teachers of African American children*. San Francisco, CA: Jossey-Bass.

Martin, L. C., & Towers, J. (2009). Improvisational coactions and the growth of collective mathematical understanding. *Research in Mathematics Education, 11*(1), 1–19.

Morgan, N., & Saxton, J. (1994). *Asking better questions: Models, techniques and classroom activities for engaging students in learning*. Markham, Ontario: Pembroke.

National Council of Teachers of Mathematics (NCTM). (2000). *Principles and standards for school mathematics*. Reston, VA: Author.

National Governors Association Center for Best Practices & Council of Chief State School Officers. (2010). *Common Core State Standards for Mathematics*. Washington, DC: Authors.

Nicol, C., & Crespo, S. (2005). Exploring mathematics in imaginative places: Rethinking what counts as meaningful contexts for learning mathematics. *School Science and Mathematics, 105*(5), 240–251.

Office for Civil Rights. (2012). *Gender equity in education: A data snapshot*. Washington, DC: U.S. Department of Education.

Schoenfeld, A. H. (1998.) Making mathematics and making pasta: From cookbook procedures to really cooking. In J. Greeno, G. Greeno, & S. V. Goldman (Eds.), *Thinking practices in mathematics and science learning* (pp. 299–319). Mahwah, NJ: Erlbaum.

Schwandt, T. A. (1994). Constructivist, interpretivist approaches to human inquiry. In N. K. Denzin & Y. S. Lincoln (Eds.), *Handbook of qualitative research* (pp. 118–137). Thousand Oaks, CA: SAGE.

Van Alsburg, C. (1984). *Mysteries of Harris Burdick*. Boston, MA: Houghton Mifflin.

Vygotsky, L. S. (1978). *Mind in society: The development of higher psychological processes*. Cambridge, MA: Harvard University Press.

Warner, C. D. (1997). The edging in of engagement: Exploring the nature of engagement in drama. *Research in Drama Education, 2*(1), 21–42.

Warner, C. D. (2007). Framed expertise: Drama as mathematical and scientific inquiry. *Research in Drama Education, 10*(15), 75–110.

Wood, T., Cobb, P., & Yackel, E. (1992). Change in learning mathematics: Change in teaching mathematics. In H. H. Marshall (Ed.), *Redefining student learning: Roots of educational change* (pp. 177–206). Norwood, NJ: Ablex.

CHAPTER 3

TECHBRIDGE GIRLS

Inspire a Girl to Change the World Through STEM

Linda S. Kekelis

ABSTRACT

Techbridge is a nonprofit organization with a mission to empower girls to achieve their goals through science, technology, engineering, and mathematics (STEM). Its after-school and summer programs offer design-based activities that make STEM accessible and engaging for girls and promote perseverance, critical thinking, teamwork, and leadership. Career exploration is embedded in activities, and role models and field trips are intended to inspire girls to pursue STEM studies and careers. Launched in 2000 and headquartered in Oakland, California, Techbridge after-school programs have served over 6,000 girls, with a particular focus on underserved girls. Techbridge supports girls in grades 4–12, creating networks that bridge elementary and middle school as well as middle and high school. This continuity of programming creates a network of peers and adults that engages girls in STEM through critical transition points when identities are shaped and academic and career goals are refined. Evaluation results attest to the program's success in increasing girls' confidence, interest, and engagement in STEM.

Out-of-School-Time STEM Programs for Females, pages 55–79
Copyright © 2017 by Information Age Publishing
55

Imagine that a young woman has the ability to develop a breakthrough for treating cystic fibrosis. Would it make its way from an idea sketched in a journal to an innovative prototype? Many opportunities await females to contribute to societal progress, and STEM (science, technology, engineering, and mathematics) can be an avenue to those opportunities. Unfortunately, many girls do not study computer science or aspire to a career in engineering because they have not had access to activities that might activate and maintain their interest.

Out-of-school-time (OST) programs can help girls to find new interests and imagine fresh possibilities for their future. Given exploration time, they might reimagine scrap materials into toys accessible for a child who is visually impaired. In a design challenge, they might learn to code an app to connect students with resources for locating internships in their community. Experiences such as these not only introduce girls to new applications for technology and engineering that they find relevant, but they can also foster confidence and a mindset that girls can learn challenging new technology skills through practice and effort. These opportunities are important for girls who might never have had a chance to try such pursuits because no one perceived—not parents, teachers, or even themselves—that they could.

For girls who do not have a home computer or a chance to explore STEM in a summer program, an after-school program can provide opportunities that could change the course of their life. These girls are at the center of Techbridge and have helped shape the program from its initial design through the new curriculum that is developed each year.

This chapter describes a multi-faceted OST program model that builds a network of STEM support for girls and provides training and resources to parents, teachers, and role models. The long-term, sustained relationships across this network are a hallmark of Techbridge, and they play important roles in broadening participation of underrepresented groups in STEM (Bell, Lewenstein, Shouse, & Feder, 2009).

BACKGROUND: UNIMAGINED POTENTIAL

Only 12% of engineers in the United States are women, and less than 3% are women from underrepresented racial/ethnic groups (National Center for Science and Engineering Statistics, 2013). For many girls, the likelihood of pursuing a STEM career is already slim by the time they enter high school (Modi, Schoenberg, & Salmond, 2012). The gender gap in high school students' interest in STEM, which has held steady the past 20 years in favor of males, is now increasing at significant rates (STEMconnector & My College Options, 2012). Research supports the importance of planning early for careers in science and that attracting students into STEM fields is best started in elementary and middle school (Maltese & Tai, 2010). Role models can help shape life choices (Dasgupta & Asgari, 2004; Evans, Whigham, & Wang, 1995), but most girls have few opportunities to see women actively involved in engineering or technology. For low-income and underrepresented girls, the chances of knowing a woman working in STEM with whom they can identify are small.

Reasons for the gender gap in STEM are multifaceted. Girls have the ability to pursue careers in these fields, but they might not have the interest, confidence, motivation, or awareness of how these fields can be rewarding (Agrawal & Donner, 2014; Eccles, 2007). Girls receive less guidance than boys about career options, less encouragement to prepare themselves for the future, and fewer opportunities to study advanced mathematics and computer science (College Board, 2013; Office for Civil Rights, 2012). A survey of parents revealed that parents are more likely to encourage boys (24%) than girls (5%) to consider an engineering career (Harris Interactive for the American Society for Quality, 2009). Further, parents were twice as likely to encourage girls to become an actress than an engineer.

Girls from communities in need face the greatest disadvantages and fewest chances to develop positive attitudes about STEM, as their communities lack resources and access to opportunities. The After-School Corporation (TASC) reports that by the time children born into poverty start their first day of middle school, they have missed out on 6,000 hours of enrichment learning activities as compared with middle-class children (TASC, 2013). This could very well mean that by her first day of sixth grade, a girl attending a Title 1 school in Oakland, California—one of the communities served by Techbridge—had never visited a science museum, received less than 20 school days of science instruction, and believes that science "isn't for girls." How many startling statistics do we need before we reimagine the STEM workforce and invest in programs and professional development to support girls' engagement in STEM?

PROGRAM DESCRIPTION

History of Techbridge

We started Techbridge in 2000 at the Chabot Space & Science Center in Oakland, California. We developed it due to concern about the lack of women in engineering and technology and the lack of promise in the pipeline, given the low number of women majoring in these fields at our local universities. We were also concerned by the number of girls participating in our summer programs at the Science Center. Girls enrolled in science classes but were not nearly as likely as boys to participate in the computer and LEGO camps. We wanted to change that.

With a planning grant from the National Science Foundation (NSF), we talked with girls in our community to determine what might interest them in engineering and technology. We conducted focus groups with girls in middle and high schools who represented the demographics of their schools in Oakland Unified School District. They articulated what they wanted: hands-on experiences, experiences different from "school science," a program just for girls, and role models to help them see possibilities.

The girls' ideas drove the design for our after-school programs, which were launched with support from the NSF. The girls-only learning environment encourages girls to try something new and to challenge gender stereotypes. Design-based activities with a hands-on approach make engineering and technology accessible and engaging for girls. Role models and career exploration activities expand how girls view themselves and their potential career paths.

Since its inception, Techbridge has served over 6,000 girls in after-school programs and has supported underserved communities across seven San Francisco Bay Area school districts. Techbridge spun off from the Chabot Space & Science Center and became an independent nonprofit in 2011, with headquarters in Oakland, California. (For more information, visit www.techbridgegirls.org.)

Techbridge supports various individuals—peers, families, teachers, and role models—in an effort to turn girls on to STEM in and out of school. With a grant from NSF, Techbridge expanded to Seattle in 2014 and to Washington, DC in 2015. By advancing effective practices and collaborating with partners, Techbridge has helped promote girls' engagement in STEM across the country.

Who We Serve

> From 6th grade to 12th grade she has transformed from shy
> and uncertain to confident, poised, and truly interested in pursuing some
> form of technology design. We credit your terrific staff, enlightening field trip
> adventures, and vision and perseverance.
>
> —Parent of Techbridge girl

Techbridge currently serves elementary, middle, and high school girls in after-school and summer programs in Oakland, San Jose, Greater Seattle, and Washington, DC. Many of these girls become the first in their family to pursue higher education. In order to maximize the number of underrepresented girls served, Techbridge partners with teachers to recruit and retain girls who are Hispanic, African American, or English language learners. Targeted and personalized outreach and support to families also promotes broader participation. The long-term, sustained relationships among peers, teachers, and staff help to broaden participation of under-represented groups in STEM.

Girls and their families complete an application, including writing an essay about why they want to participate in Techbridge. In most schools, the number of girls who want to attend can be accommodated. In schools where demand exceeds capacity, girls are chosen by lottery with wait lists maintained. Girls make a commitment for the academic year and engage in approximately 75 hours of programming annually. Many girls return for a second year and some participate in Techbridge up to seven years. Programs are held weekly at schools during after-school hours. This helps make the programs more accessible to girls and reduces transportation challenges and costs. The girls walk home, take public transportation, or are picked up by their parents afterwards. Some schools schedule a bus to pick up students who participate in Techbridge and other after-school programs. We find it easier for working families to support their daughters' participation when Techbridge coordinates with other after-school programs at their children's schools.

By supporting girls from the fourth through twelfth grades, Techbridge embodies its name by serving as a bridge to introduce girls to projects and people that can spark an interest in STEM, and over time, links girls with a support system that can sustain their identity in STEM. For older girls, the program introduces resources that offer access to social capital (helpful social connections) and supports their academic and career journeys in STEM (Stanton-Salazar, 2011).

Techbridge's Proven Model: The Secret Sauce for Inspiring Girls

The "heart" of Techbridge is after-school programs where girls experience STEM through hands-on projects. During the program, the girls build confidence and find that they like to do metal work (soldering), tinker with tools, design video games, and dismantle appliances to learn how they work. We look to research and to our girls for inspiration to choose and design projects that appeal to girls. The projects inspire the girls to ask questions and want to learn more. They promote inquiry and incorporate real-world STEM applications.

Techbridge stays abreast of girls' interests through observations, surveys, and focus groups and uses this input to design curriculum. The curriculum has been developed, evaluated, and refined over 16 years. Techbridge curriculum is piloted with girls in our Summer Academy, reviewed by educators, and revised over multiple cycles. We solicit input from a range of content specialists, including scientists, engineers, teachers, and external experts.

Techbridge offers a wide range of curricula, from engineering design challenges that require inexpensive and simple materials such as newspaper, popsicle sticks, and recycled appliances to advanced technology projects that use computers and hardware, including Arduinos, electronic sensors, and 3D printers. Many projects are design-based and focus on topics relevant to girls, difficult enough to be interesting and engaging enough to promote career interests.

We often hear from girls that they want to make the world a better place but do not realize that STEM careers align with this aspiration (Cohoon & Aspray, 2006). Part of what makes Techbridge projects engaging for girls is that they include a real-world component. For example, in *Girls Go Global*, girls design and build their own solutions to problems, including biomass-burning stoves and LED lanterns. They design a device to carry water and in the process learn how to work in teams and to understand the experiences and needs of girls in other parts of the world.

The curriculum introduces girls to a variety of engineering and technology fields, with a focus on disciplines where women are underrepresented, including electrical engineering and computer science. For example, in *Tinkerbelles and Tech Ninjas* girls learn electrical engineering and computer science as they modify a personal object of their choice to solve a problem in a creative way. They work with LEDs and Lilypad Arduinos to make wearable technology, including t-shirts, hats, and pillows. In *Product Design*, the girls analyze existing designs by taking apart household appliances, such as hairdryers, toasters, and game console controllers, and apply this analysis to designing their own toys. In *Green Design*, the girls create green dollhouses using recycled materials and are led through a simulated LEED certification by role models from green construction. As the girls work through these units, they engage in the practices of engineers and scientists, acquire core STEM knowledge, and apply interdisciplinary concepts.

Grit: Celebrating Glorious Goofs

In our programs we place on the wall a poster for Celebrating Glorious Goofs, where girls and staff step back from their mistakes and reflect. They write how they feel and what they learn for everyone to see. This activity is a positive outlet for the challenges likely to be experienced in technology and engineering projects (Kekelis & Ryoo, 2015).

Grit and perseverance are important life skills. They are essential for engineers and scientists who work on projects that have no easy solutions. They support students en route to completing a computer science degree. Techbridge helps girls build their "confidence muscle" by introducing projects that are fun *and hard*. Many projects involve the engineering design process; its do-overs give girls a chance to gain confidence and experience how they can work through problems.

We challenge our high school girls with complex projects, such as building an underwater vehicle that girls operate remotely. Working with PVC piping and motors, girls learn about electronics and buoyancy. Just as important as successfully completing these projects is learning how to persevere and not give up the first or second or third time. Experience with troubleshooting and redesign helps girls develop the mindset that failure is an important part of learning. The girls experience and celebrate hard work through trial and error and learn when to go it alone and when to enlist support from others. These life lessons will serve them well when they encounter challenges in their studies and on the job. We hear from alumnae that they draw on these Techbridge lessons when they face challenges in their college classes.

Sisterhood: Building Community and Learning Teamwork

Techbridge isn't a program—it's a family that cheers when you are successful and is there when the road gets rough.
—Techbridge student

The familiar staff and participants in Techbridge bring comfort and predictability that girls welcome, as they are faced with change and uncertainty in their lives. The girls say that Techbridge feels like a family or a sisterhood. One teacher reflected on this climate by stating, "The supportive, encouraging nature of the Techbridge program serves as a much-needed antidote to the pressure many adolescent girls feel to 'be cool,' which can often mean anything but excelling at science and technology" (Furger, 2002). This type of peer pressure as experienced in Techbridge serves a favorable role. Students' reflections in journal entries and end-of-year surveys reveal that these group experiences, which are highly affirming, foster a positive learning environment and reinforce positive feelings towards STEM.

Warm, trusting environments take time to develop and hard work to sustain (Kekelis, Ancheta, Heber, & Countryman, 2005). Essential to Techbridge is building community throughout each session and over time. Icebreakers and check-in circles at the start of each session support relationships across girls. The girls find common bonds that bridge grade levels, race/ethnicity, and friendship groups. These positive relationships allow girls to brainstorm without fear of being ridiculed, to gracefully respond to failed task attempts, and to practice public speaking.

Building on research, we make projects collaborative in a variety of ways (Fancsali, 2002; Regalla, 2010). In *pair programming*, the girls work with a partner on computer science projects; this collaborative, instructional approach increases learning and interest in computer science (Werner & Denner, 2009). We use *pair partnering* when the girls work on projects of their own but in the company of a partner to whom they can turn to give or get help. This arrangement builds confidence and sustains engagement.

We are intentional in reinforcing teamwork. When girls finish a task early, they are encouraged to see if anyone needs assistance. When in need of help, we have a mantra "ask three, then me," which reminds the girls to solicit support from peers before adults. At the end of each session, the girls and adults share shout-outs that often recognize the efforts of teamwork. In their journal reflections, the girls write about their experiences in teams, both in terms of their contributions as a team member and how they promote teammates' participation.

The Value of Empowerment: Building Leadership

Not every Techbridge girl will follow a career path in technology or engineering, but every Techbridge girl will need to advocate for herself, know how to network, and feel confident at public speaking. Techbridge provides opportunities to practice these skills over time and across settings. For example, we learned that girls need guidance and encouragement to make the most of meeting with role models, so we developed strategies to support girls' success. These include establishing an expectation that every girl will ask at least one question during a role model visit and helping girls practice introducing themselves to role models on field trips.

Because public speaking is an important skill, as well as a challenge that most girls fear, we devote resources that support girls' growth in this area. Icebreakers, such as *Don't Say Um*, give girls experience making presentations with fluency and without fillers such as "um" and "you know." From the relative comfort of participating in a pair-share task to explaining a project to the entire Techbridge group, the girls gain greater comfort exercising their voice. These activities build toward presentations made at Techbridge family nights. While these experiences make the girls uncomfortable at the time, our alumnae express appreciation for us having nudged them to go beyond their comfort zone and practice public speaking. We have heard from parents, teachers, and the girls that the confidence the girls developed helped them assume more leadership roles, such as making a speech at their graduation and being more engaged in science class discussions.

We look for opportunities beyond Techbridge for the girls to demonstrate leadership. Within the program, older girls take on leadership and lead hands-on activities and help host family events. High school girls serve as role models and pay it forward, returning to middle schools and offering academic guidance and insights into what to expect in high school. We also invite girls to participate on panels at conferences, write blogs, and apply for awards.

Career Exploration: Role Models Expanding Options

STEM activities, no matter how engaging, might not generate career interests. In our first year, girls expressed that while they loved hands-on projects, they viewed them as hobbies and not as career options (Kekelis, Ancheta, & Heber, 2005). They also stated that while adults encouraged them to follow their dreams, they wanted more specific academic and career guidance. We turned to role models and field trips to transform activities into career possibilities and help fill the "guidance gap." While we do not expect that every Techbridge girl will become an engineer or computer scientist, we do think it's important to show the possibilities in engineering and technology that girls might have never imagined.

Our after-school programs host at least two visits by role models and two field trips to work sites and universities. Techbridge has partnered with more than 1,000 role models from technology, science, and engineering firms. During visits to our after-school programs, role models engage in STEM activities with the girls. During field trips to work sites and universities, the girls see first-hand what engineers and scientists do, and they interact with a variety of role models. The role models share information about their studies and work and help the girls make connections between the Techbridge activities and careers, as well as to understand the positive impact they might have one day as engineers or computer scientists.

We also integrate interactive career exploration activities throughout our programs. From designing a career card to watching a video of an engineer in the field to assuming the role of a chemical engineer, the girls are invited to make ongoing connections between hands-on activities and STEM careers.

Training: Setting Role Models Up for Success

Knowing how to lead an activity or talk about careers in a relevant way for students is not something that scientists or engineers learn in college or on the job. Without support, well-intentioned role models might fail to connect with students because they do not know how to communicate their passion for their work or how to make a presentation that is appropriate for their audience. Techbridge allays fears and sets role models up for success by training them (Kekelis & Joyce, 2014).

Techbridge identified the following elements for role models' success: (a) being personal and passionate about their career; (b) introducing engaging, interactive activities that relate to work; (c) explaining why STEM matters; (d) dispelling stereotypes by sharing hobbies and social pursuits; and (e) providing academic and career guidance. The key message presented to role models is: It is personal stories and interactions that help students relate to role models and see STEM in a favorable light (Techbridge, 2013).

Techbridge program coordinators meet with and offer support to role models before classroom visits and field trips. They provide background information on the girls, help identify activities to lead, and coach them on how to share personal stories. The program coordinators follow up after role model visits and field trips to celebrate the successes as well as debrief and explore how Techbridge and the partners can improve upon their efforts.

Techbridge developed role model guides and online resources that include icebreakers, hands-on activities, tips for company tours, and questions to promote interactions between girls and role models. These resources can be used across all STEM disciplines and profile case studies of role models in a variety of fields.

Family Engagement: Bringing STEM Programming Home

During our *Cars and Engines* Summer Academy, a father commented that he had never thought to invite his daughter to help him work on the family car. This was an experience that he and his son enjoyed doing together. During our summer program, when his daughter expressed interest in and knowledge about cars, he was surprised and promised to include her in future repairs to the family car. Parents often do not realize that their daughters might like to tinker, work with tools, or take on a household repair project, and girls themselves might not ask to engage in these projects.

We want all parents to feel capable of supporting their daughters in STEM. We heard from one father that he felt inadequate because he had not been to college and was not fluent in English. This father took time off from work and rearranged his schedule so that he could drive his daughter to weekend Techbridge events. Encouragement from family, regardless of their technical expertise, can foster and reinforce interest (Google, 2014).

Our after-school and summer programs offer family events, which include meals and childcare, and provide an opportunity for parents to celebrate their daughters' accomplishments and learn how they can support their daughters' interests in STEM (Rosenberg, Wilkes, & Harris, 2014). By seeing their daughters engaged in and excited about learning, parents discover that engineering and technology are fields of study in which their daughters can thrive.

We share resources with families throughout the year. Gift ideas are sent home before the holidays. These include toys and games that promote STEM learning, offerings at local museums, and projects that can be completed with simple household items. Summer opportunities for science and technology programs, with scholarship information, are mailed home. We also share research with families. For example, we highlight a growth mindset and encourage parents to be mindful of the messages they share and to encourage effort and learning.

We offer at-home STEM projects online and in print. *Science: It's a Family Affair*, a guide developed by Techbridge, provides resources for engaging girls in STEM careers (Pena, Kekelis, Anaya, & Joyce, 2013). It is available in English, Spanish, and Chinese. We also use text messaging and phone calls to connect to families and share highlights of local resources.

Collaborative Relationships

Partnerships are key to our success. Techbridge partners with role models at organizations and universities, including Chevron, Google, Samsung, Cisco, Salesforce, Oracle, Microsoft, Stanford University, and the University of California, Berkeley. We strive to recruit role models who are from the community and represent our girls. These individuals show that academic and career success in STEM is possible and highlight STEM career paths. We are explicit in our partnership requests, seeking partnerships with professional organizations such as the Society of Hispanic Professional Engineers and the National Society of Black Engineers to recruit a diverse group of role models.

Partners introduce our girls to a wealth of opportunities, helping create the ecosystem that puts youth at the center and supports them over time and across settings (Traphangen & Traill, 2014). Opportunities include the Smith Summer Science and Engineering program, the Berkeley Foundation for Opportunities in Information Technology, and the Aspirations in Computing Award offered by the National Center for Women and Information Technology. These partners help the girls gain valuable experience learning how to access academic support, write essays for scholarships, and apply for awards.

Techbridge also uses the resources and lessons learned from our after-school programs to support partners, building their capacity to offer engaging STEM activities to more girls. Techbridge provides training and technical assistance on our curriculum. We work with a growing list of partners, including school districts, Boys & Girls Clubs, YMCA of the USA, and others. Through Girls Go Techbridge, Techbridge delivers curricula and career resources to Girl Scout councils across the country. Techbridge partners with the Society of Women Engineers, the National Girls Collaborative Project, and Girl Scout councils through the Role Models Matter project, providing resources and training for role models across the country.

Staff

Techbridge provides high-quality programs by staffing after-school programs with a program coordinator in partnership with a classroom teacher. Together, program coordinators and teachers plan and host the after-school programs. They work together in each session, which allows for a ratio of 2 adults to 20–30 girls. For younger girls and for more complicated technology projects, this level of support proves particularly helpful in keeping girls engaged and projects progressing. As trusted adults and young and dynamic role models, program coordinators provide expertise in STEM and youth development and mentor the girls. These coordinators work full-time and prepare materials, train role models, plan field trips, pilot curriculum, and support Techbridge in outreach events with partners in the community. The majority of our program coordinators have experience in STEM and teaching in and/or out of school. Classroom teachers co-lead programs and recruit a broad range of girls. They help the girls relate what they learn in the program to what they are learning in the classroom. We recruit teachers in consultation with principals and look for educators who possess passion for STEM programs for girls and with whom girls can relate well. Teachers receive a stipend for their engagement in the program and associated professional development activity as noted below.

Ongoing training is offered to program coordinators and teachers. Quarterly meetings, which are planned and hosted by Techbridge managers and staff, provide time to practice new curriculum and learn how to introduce it to girls. These meetings offer the added benefit of allowing teachers to network across schools and share lessons learned. Program coordinators meet weekly to learn from one another and engage in professional development by participating in conferences and webinars, meeting with peers and colleagues, and staying current with the research and practice literature.

Funding

Techbridge after-school programs are free for participating girls, the majority of which qualify for federal free or reduced-price meals. The program was launched with support from the NSF and has been sustained through funds from participating schools and a variety of sources that include corporations, foundations, and individual donors.

PROGRAM EFFECTIVENESS

Techbridge uses a mixed-methods, quasi-experimental evaluation plan to assess curriculum and program outcomes through pre- and post-surveys, interviews, focus groups, and observations. Participants involved in one or more of these methods include girls, teachers, role models, and family members. Data from interviews, focus groups, and observations of girls, teachers, and role models are collected throughout the year and are used by managers to inform the development of curricula and the program model. The external evaluation of the program is conducted by Education Development Center (EDC). Highlights of their findings from surveys conducted with girls, parents, and teachers for San Francisco Bay Area programs in the 2013–2014 school year (Fitzhugh, 2014) are presented in this section.

Student Sample

In this chapter a subset of results, retrospective items on the post-survey, are presented for the Techbridge girls. These girls participated in the 15 San Francisco Bay Area schools in Oakland, San Lorenzo, and San Jose. A total of 344 Techbridge girls completed the post-surveys. Of these, 28.5% were in elementary school (grade 5), 56.5% were in middle school (grades 6–8), and 15.0% attended high school (grades 9–12). The self-reported race/ethnicity of these girls was 47.4% Hispanic/Latina, 26.1% Asian, 9.9% Black/African American, 9.9% Multi-racial, 5.7% White, 0.6% Pacific Islander, and 0.3% American Indian.

Data Collection and Analysis

Techbridge students were given pre-surveys in the fall, shortly after the start of the after-school programs and in the spring, shortly before the end of the programs. Student surveys included 14 items from assessments with validated scales that measure knowledge, confidence, attitudes, and interest in STEM; career plans; and engagement in STEM-related activities outside of school and Techbridge. The girls were asked to give the program a letter grade and explain their rating. The Techbridge girls completed a retrospective survey at the end of the program year in which they indicated their level of agreement with 17 items that address Techbridge's impact on their career aspirations, self-confidence, technical abilities, and scientific mindset. An Analysis of Variance (ANOVA) was conducted at $\alpha = .05$ significance level on retrospective outcomes across race/ethnicity, grade level, and participation time in Techbridge. In this chapter only the results from the retrospective items on the post-survey are presented from the Techbridge girls due to space constraints. (Contact info@techbridgegirls.org to learn more about the surveys and other program-evaluation instruments.)

Parents of Techbridge girls were given surveys to complete during family events in the spring. For parents who did not complete surveys at these events, surveys were mailed home. Surveys were available in English, Spanish, and Chinese. A total of 180 parent surveys were completed, which represents 52% of the girls who completed a post survey. The survey includes seven closed-format questions consisting of four questions about parents' perceptions of the program's influence on their daughters and three questions about perceived influence on them. Sample questions include: "Because of Techbridge I am more aware of activities in science, technology and/or engineering that my daughter can participate in (like museums, camps, classes, etc.)" and "Because of Techbridge I encourage my daughter to participate in more technology and science activities (taking advanced math and science classes, going to a science museum, attending summer programs, building things at home, etc.)." In addition, five open-ended questions invited input on specific resources for families, such as the Techbridge Family Night and newsletter, along with opportunities to explain how Techbridge made a difference in their family and suggestions for improving the program. The data are presented as percentages for the closed-format questions and narrative description for open-ended questions.

All 18 teachers who co-hosted programs during the 2013–2014 school year completed an online survey at the end of the school year. This survey included 13 closed-ended questions about the teachers' perceptions of the program's impact on girls and seven closed-end questions about the program's impact on themselves. Response choices included: not at all; to a small extent; to some extent; to a large extent; and to a very large extent. The survey also presented open-ended questions in which teachers could describe the impact of field trips and role model visits on girls, as well as offer suggestions for changing these program elements. Results are presented as percentages for the closed-ended questions and narrative description for main themes appearing in the open-ended questions.

Findings

Techbridge's Influence on Girls

The girls reported increased STEM knowledge and skills, improved problem-solving skills and perseverance in STEM, better public speaking skills, and increased knowledge of and more favorable dispositions toward STEM careers. In this chapter, a subset of the wider data is presented. Highlights from results of the 2013–2014 post-survey items completed by 344 Techbridge girls along with results from the teacher and parent surveys are presented below.

Improved technical and scientific knowledge and skills. Girls, teachers, and parents all reported that Techbridge helped participating girls improve their technical and scientific knowledge and skills. Two retrospective questions on the student post-survey asked Techbridge participants to indicate whether the program influenced their science, engineering, and technology skills. Most girls indicated that Techbridge had favorably influenced their knowledge of basic physics concepts and using computer programs. Approximately 88% of Techbridge girls said they know more about how things work (including circuits and simple machines), and about 79% said they are better at using new computer programs.

Similar to the girls' survey results, teachers reported that the girls increased their knowledge about basic physics concepts and computer programming, more so for the former than the latter. While 89% of teachers reported that girls knew more about how things work either *to a very large* or *large extent* because of Techbridge, 53% of teachers reported that the girls knew more about computer programming *to a very large* or *large extent* because of Techbridge. In response to an open-ended question on the teacher survey asking teachers in what key areas Techbridge had the greatest influence, 2 of the 18 teachers (11%) noted that the girls had increased their STEM skills. One said, "They understand more about how things work."

Parents also said their daughters' knowledge and performance in STEM improved. In response to an open-ended question regarding Techbridge's influence, about one in five parents (22%) said their daughter's knowledge and skills in science and/or technology improved. One parent commented, "She's gained knowledge in computers and programming. Working with coding has widened her work skills."

Most parents reported that Techbridge had helped their daughters improve their mathematics and/or science grades, with 92% agreeing or strongly agreeing. In the open-ended question regarding Techbridge's impact, some parents (6%) said their daughter had become more engaged in school and/or were doing better in school because of Techbridge. Sample parent comments include: "Her grades in school have improved" and "My daughter has a lot more enthusiasm about school and science."

Increased confidence in science, technology, and engineering. Girls, teachers, and parents all reported that Techbridge helped girls improve their confidence in science, engineering, and technology. Two questions on the student post-survey asked students to indicate whether Techbridge had influenced their STEM confidence. The majority of Techbridge students reported that Techbridge helped them become more confident in using technology (84.5%) and in science (79.5%).

The teachers and parents also reported that the girls' confidence in science, engineering, and/or technology had increased as a result of Techbridge. Almost 72% of the teachers indicated that girls' confidence grew *to a very large* or *large extent*, while the remaining 28% said that girls' confidence grew *to some extent*. In response to an open-ended question on the teacher survey asking teachers in what key areas Techbridge had the greatest influence on girls, four teachers (22%) said they thought Techbridge had favorably influenced girls' confidence in STEM. One teacher wrote, "I do believe that the girls are much more confident and able to take risks than before the program."

Most parents (98%) agreed that Techbridge helped their daughter become more confident in science, technology, and engineering, including 54% who strongly agreed that their child's confidence increased because of Techbridge. In an open-ended question where parents were invited to describe how Techbridge had influenced their daughter, about one in six parents (16%) indicated that their daughter's confidence in science, engineering, and/or technology improved. One parent commented, "[My daughter] is more confident and outgoing when talking about science and technology. She is also more relaxed when going to a new class or event." Another parent stated, "I knew that my daughter was smart, but now she is more confident and secure."

Enjoyment of science, engineering, and technology activities. Most Techbridge students (92%) agreed that it was fun to learn in Techbridge. Techbridge participants were asked to grade Techbridge on a five-point scale from A to F. The majority of girls gave Techbridge an A or B (88%), with 57% of them giving the program an A. The girls explained that they rated Techbridge highly because they had an opportunity to learn and improve their skills in STEM specifically (20%) or more generally (21%); they had an opportunity to do fun, hands-on projects (14%); they gained confidence (5%); they formed a positive relationship with their Techbridge program coordinator, teacher, and/or role models (4%); and/or they learned that girls can succeed in STEM (4%). One girl wrote, "I gave Techbridge this grade because the teachers here help me understand more things and it's really fun at Techbridge. The teachers make me laugh and it makes me fit in here."

The girls were asked to explain what they liked most about Techbridge. The most frequently cited reason—given by more than half of the girls

(61%)—was doing hands-on projects. Several said Techbridge gave them the opportunity to do what scientists and engineers do. One commented, "What I liked most about Techbridge was the activities we had and how we struggled to solve them."

The girls also said they particularly enjoyed the field trips (26%); the social aspects of Techbridge (16%); learning in general (10%); and the Techbridge staff, teachers, and/or role models (8%). A few girls commented that they appreciated the girls-only environment. One said, "Techbridge is made for girls and I feel that its environment is easier and more friendly to work in."

Improved problem-solving skills in science, engineering, and technology and greater perseverance in overcoming obstacles in science, engineering, and technology. Techbridge girls, teachers, and parents reported that Techbridge helped the girls become better problem solvers and persevere in the face of obstacles. Three questions on the post-survey asked the girls whether Techbridge had improved their problem-solving skills and perseverance, which are closely related skills and habits. A total of 92% of girls agreed that it can take many tries to solve a problem (including 50% who *strongly agreed*), and 84.5% of girls agreed that they now try harder to overcome a challenge (including 41% who *strongly agreed*).

The teachers agreed that the girls were better problem solvers because they participated in Techbridge. They were especially likely to report that Techbridge's girls-only environment helped girls feel freer to experiment. Most teachers (89%) said the girls were more willing to try new science and technology activities either *to a very large extent* or *large extent* because Techbridge only serves girls. The majority of teachers also reported that Techbridge helped the girls better understand that it can take many tries to solve a problem *to a very large* or *large extent* (82%), become better at problem solving *to a very large* or *large extent* (78%), and try harder to overcome a challenge *to a very large* or *large extent* (72%).

In response to an open-ended question on the teacher survey asking in what key areas Techbridge had the greatest influence on the girls, teachers were most likely to report that Techbridge had increased the girls' perseverance and problem-solving skills. Ten of the 18 teachers (56%) reported that Techbridge had favorably influenced girls' grit. One teacher wrote, "Girls are more willing to take risks. Many have shown an impressive quantity of resourcefulness in approaching challenges and determination in the face of setbacks."

Parents were not asked specifically about improvement in their daughters' problem-solving skills, but several parents commented in response to an open-ended question about Techbridge's influence that the program had helped their daughters become more creative and persistent problem-solvers. One said, "I see she's exploring more and not giving up in just one try."

Increased knowledge about science, engineering, and technology careers.
One of Techbridge's goals is to help girls understand various career options
in science, engineering, and technology and to begin to learn about the
pathways toward these careers. Most Techbridge girls (87%) *strongly agreed*
or *agreed* on the post-survey that Techbridge increased their knowledge
about various careers. The program appeared to most strongly influence
the girls' knowledge about engineering and technology careers.

The teachers also thought Techbridge helped girls become more knowledgeable about STEM careers and career pathways. Most reported that the
girls were more knowledgeable about what STEM workers do *to a large* or
a very large extent (83%). The teachers were also likely, but to a somewhat
lesser degree, to report that the girls had become more knowledgeable
about STEM career pathways, with 69% of teachers reporting that the girls
had increased their knowledge about STEM career pathways *to a large* or *a
very large extent*.

A few parents responded to an open-ended question that Techbridge
increased their daughters' knowledge about STEM careers. One wrote,
"[Techbridge] opened her mind to new opportunities available to women."

Better dispositions toward science, engineering, and technology careers.
Two post-survey questions assessed girls' interest in working in technology,
science, or engineering careers. While approximately 82% of Techbridge
girls said Techbridge made them more interested in STEM careers, 69%
said they could actually see themselves working in STEM.

In the open-ended question where parents were invited to describe how
Techbridge had influenced their daughter, several parents (13%) indicated that their daughter was more aware of (and often more interested in)
various career options and pathways. For example, one parent said her/his
daughter was "more motivated to look into careers involving engineering
and technology."

Better public speaking skills. The girls and teachers were asked whether
the girls had improved their public speaking skills because of Techbridge.
The girls reported relatively less influence on these skills than other program outcomes, with 62% of the girls agreeing that Techbridge helped
them become more comfortable speaking in front of a group of people
(including 24% who *strongly agreed*).

Teachers were also somewhat less likely to report that the girls had improved their public speaking skills as compared with other student outcomes.
Half the teachers (50%) indicated that the girls were more comfortable
speaking in front of a group of people *to a very large* or *large extent*, while 39%
of the teachers indicated that the girls were more comfortable *to some extent*
and 11% reported the girls were more comfortable *to a small extent*.

A few parents commented on the survey that their daughters had become more assertive and better public speakers. One stated, "She is much

more confident when she thinks of a hypothesis or solution. She speaks out more with more confidence." Although evidence indicates improvement in the girls' speaking skills overall, this is an area we are working to strengthen.

Techbridge's impact on families.   As a result of Techbridge, parents reported that they were more aware of resources and encouraged their daughters' engagement in STEM. The resources helped keep parents informed on the projects in which their daughters were engaged and thereby promoted conversation at home. One family described how they were able to relate what's learned in Techbridge to real-world applications. They also accessed additional opportunities in the community, as noted by a parent: "I used the resources and found a program for Latinas in STEM. We went to a conference and were so encouraged." Ninety-seven percent reported that they agreed (with 63% strongly agreeing) that they encouraged their daughter to participate in more technology and science activities as a result of Techbridge. Ninety-seven percent agreed (with 58% strongly agreeing) that they were more aware of STEM activities that their daughter can participate in. Ninety-four percent agreed (with 53% strongly agreeing) that they encouraged their daughter to pursue careers in science, technology, and engineering.

Families also offered suggestions for program improvement to better support family engagement. These included filming field trips so that families could see what their daughters were learning, inviting families to programs earlier in the school year and more often, and translating materials into Vietnamese. We are exploring ways to respond to families' interests and needs.

Techbridge's impact on teachers.   The program showed success in building confidence and capacity for participating teachers as follows: 55% reported increased ability to engage girls in STEM projects *to a very large* or *large extent;* 59% reported increased knowledge and awareness of STEM careers *to a very large* or *large extent;* 59% reported increased knowledge of other science and technology resources and programs *to a very large* or *large extent;* and 33% reported increased technical skills *to a very large* or *large extent.*

In addition, Techbridge had a multiplying effect and influenced teachers' classroom practices as evidenced by the following teacher descriptions of use of Techbridge resources during the school day:

- "I incorporate more cooperative and hands-on learning in my regular class, run less scripted, more open-ended lessons, because of what I've learned about student learning from teaching Techbridge."
- "I've become so excited about STEM careers and the maker movement that I've integrated a lot of this into my math and science curriculum— so much so that when my math and science students were asked by guests to our class about what they'd like to be when they grow up, a majority of them (both boys and girls) gave STEM careers!"

CHALLENGES AND LESSONS LEARNED

Our programs are not without challenges, which we consider opportunities to gather input from girls, families, and partners who support our programs. One key insight from our recent work is that girls look to their parents for career guidance more often than we had previously thought. We also learned that parents believe that their middle-school-aged daughters are too young for career counseling, and many parents hold beliefs that they are not qualified to advise their daughters. Without a college education, career in STEM, or fluency in English, some of our parents have said that they do not know how to support their daughters' interest in STEM. Encouragement from family members, regardless of their technical expertise, can foster and reinforce girls' interest in STEM. Techbridge is currently working on a research project to develop a model of family engagement that supports career outcomes for girls, particularly those in under-resourced communities. This research project brings families to the table to design programs that are successful in empowering parents in a space where the parents feel comfortable. One goal is to provide families with a supportive space to interact about STEM projects to spark interest and connect STEM with the home. Another is to identify strategies for making resources accessible to families from diverse linguistic and educational backgrounds. The intent is to engage with families to validate their knowledge and resources that can support girls' success in STEM studies and career pathways. Our research findings and resources will be disseminated to enhance the efforts of schools, after-school programs, and community partners to support girls in STEM. By building partnerships between programs and families, we aim to strengthen the STEM learning ecosystem.

PRACTICAL IMPLICATIONS

Since it began in 2000, Techbridge has refined its program model that demonstrates success in engaging girls in STEM experiences, encouraging girls in careers in technical fields, and promoting leadership and perseverance skills. Along the way, there have been many successes and challenges that resulted in lessons learned. Effective practices that Techbridge has developed for engaging girls in STEM follow.

1. *Make space and programs just for girls.* The girls-only element helps build confidence and is especially important in subjects such as technology and engineering, in which girls might have less exposure than boys. Without fear of being teased for not knowing how to use a power tool or how to debug a computer program, girls are more in-

clined to try new skills and persevere through failures. It is important to be explicit with girls and families regarding why the program is dedicated to girls. Statistics can help make the case for the need for more females in STEM. With girls, this discussion can be a learning opportunity to explore stereotype threat and growth mindset.

2. *Remain ever mindful of promoting equity across all groups.* Our programs bring a lot of diversity in terms of ability, experience level, social class, and race/ethnicity. It is a never-ending challenge (and opportunity) to address social dynamics such that every girl feels the right balance of being challenged and supported. While the girls like working with friends, they also express desire to go outside their friendship circles. Social engineering helps mix groups up. We have not always gotten this right and have learned when to assign partners (or ask girls to find a new partner) and when to allow girls to work with friends (Kekelis et al., 2005). We also learned to be more explicit about *why* we do social engineering and explain our reasons to the girls and families.

3. *Involve girls in program design and evaluation.* Evaluation may not be the most exciting part of programs, but we find it important to talk about why we collect data and how youth input is important to our programs. This generates new ideas to freshen programming and identifies activities that are relevant to the girls. This practice helps girls understand how to think critically about program design and improvement and how to offer feedback. Their input into survey design can also lead to items that tap into the kind of information we seek.

4. *Commit to serve all girls in STEM.* Our teachers help recruit a wide range of girls, including those with an interest in STEM and those who are less certain or confident. We encourage them to recruit girls who might struggle or hold back in class, who might engage and be successful in an after-school program. Teachers invite girls to try the program. Getting the girls in gives us a chance to dispel stereotypes about what they think a STEM program might be about and to offer them engaging activities.

5. *Explore STEM careers.* Girls receive messages from the media, friends, and families that influence how they view their future. We want to make sure that STEM is an available option for girls. From the group icebreaker to project completion and reflection, Techbridge invites girls to envision themselves holding real jobs as chemical engineers, forensic scientists, or computer programmers. Role models help girls make connections between these careers and the Techbridge activities in which they engage. Messages from the role models and staff continually reinforce the idea that girls

do science and engineering and that STEM careers are creative, relevant, and interesting.

6. *Involve families.* The majority of our girls' families participate in family events. Personal invitations prove successful in making every parent feel welcome and important. We translate materials and presentations to help make messages accessible and families valued. At every opportunity we reinforce how important families are for the girls' engagement in STEM. We strive to send all parents the message that their *encouragement* is what matters, regardless of their STEM expertise.

7. *Create an ecosystem to encourage girls in STEM.* We cannot be all things for all girls all the time. Our success depends on the network of support we create over time. This includes peers, families, teachers, role models, and partnerships with STEM organizations. We look for partners who can help us reach our mission to inspire girls and expand their future options in STEM. Sometimes partner support is concurrent with the girls' participation in Techbridge, whereas at other times this support starts after the girls have completed our program. Together, we try to help the girls maintain interest and find support to be college- and career-ready.

8. *Structure feedback and assessment into program operations.* At Techbridge we expect all staff to offer feedback on what is working well and areas that need improvement. We also offer feedback to partners, including role models. While it might be hard at first, we find it important to give and receive feedback on visits with role models. It helps to set an expectation from the start that follow-up meetings will be scheduled to share feedback for program improvement. Starting these conversations by asking partners for help on what they think can be improved makes it easier to share feedback. Everyone wants to help when they are asked. Quality feedback makes for lasting partnerships. With practice and training it gets easier to provide feedback and is important for program improvement. Evaluation results, such as those shared in this chapter, also inform the program in ways that allow research and practice to be integrated in a continuous cycle for program improvement.

For information on training, curriculum, and partnership opportunities with Techbridge, contact info@techbridgegirls.org.

CLOSING COMMENTS

By improving girls' access to STEM education and role models, we can plant seeds for the next generation of innovators—the girl who discovers the cure for diabetes or develops a new technology to assist individuals with visual impairments. After-school programs can provide girls with opportunities to experience STEM in ways that show creative, real-world applications and allow them to reimagine their identities. These opportunities are particularly important for girls who are most underrepresented in STEM, providing resources that build their assets and expand their career options.

We advocate for continued support for OST STEM programs that serve girls, as well as for research on their long-term impact. Longitudinal studies can help determine the influence of participation in OST STEM programs at various points in girls' lives. While the end goal of some of these programs is to increase the number of girls pursuing STEM careers, girls pass through a number of milestones and STEM-related choices before reaching that stage. Investment in research can help us better understand how informal STEM programs generate initial interest in STEM and how they can help girls maintain their interest and continue on an academic and career path in STEM.

It is especially important to study how to effectively implement STEM programs in communities that are under-resourced. Researching how girls from families of various cultural and socioeconomic backgrounds navigate their path toward a STEM career can provide insight into the needs as well as the assets of girls from various types of communities. Understanding how underrepresented girls—including racial/ethnic minorities, low-income girls, English language learners, and girls with disabilities—who participate in informal STEM programs are able to defy the statistics and pursue a STEM career will inform the field about how to best engage youth of all backgrounds. What are the barriers in recruiting and retaining different groups of youth in various types of STEM programs? What strategies help overcome these challenges? Research on these questions will help inform the field of informal STEM education about how to best support all girls.

We advocate for a perspective that looks at the assets of girls and families and propose the following questions: How do youth create social capital to overcome the barriers that under-resourced communities can pose? What allows some girls to use OST programs as a springboard to access STEM-related resources and to advocate for their own personal development? How do parents help their daughters maximize their social capital, and what encourages or limits their involvement? Investigating these questions can inform STEM programs how to better equip families to support girls in STEM.

The existing patchwork of after-school STEM programming might not seamlessly serve girls. Greater coordination across programs could create the ecosystem to help girls envision and explore a career path, develop grit needed for a career in STEM, and garner support along the way. We advocate for research and practice to advance coordinated programming that puts girls at the center.

REFERENCES

Agrawal, N., & Donner, J. (2014). *Frontiers in Urban Science Exploration: Resource guide* (2nd ed.). New York, NY: The After-School Corporation.

Bell, P., Lewenstein, B., Shouse, A. W., & Feder, M. (Eds). (2009). *Learning science in informal environments*. Washington, DC: The National Academies Press.

Cohoon, J., & Aspray, W. (Eds.). (2006). *Women and information technology. Research on underrepresentation*. Cambridge, MA: MIT Press.

College Board. (2013). Advanced Placement program participation. Retrieved from http://research.collegeboard.org/programs/ap/data/archived/2013

Dasgupta, N., & Asgari, S. (2004). Seeing is believing: Exposure to counterstereotypic women leaders and its effect on the malleability of automatic gender stereotyping. *Journal of Experimental Social Psychology, 40*(5), 642–658.

Eccles, J. S. (2007). Where are all the women? Gender differences in participation in physical science and engineering. In S. J. Ceci & W. M. Williams (Eds.), *Why aren't more women in science?* (pp. 199–210). Washington, DC: American Psychological Association.

Evans, M. A., Whigham, M., & Wang, M. C. (1995). The effect of a role model project upon the attitudes of ninth-grade science students. *Journal of Research in Science Teaching, 32*(2), 195–204.

Fancsali, C. (2002). What we know about girls, STEM, and afterschool programs: A summary. *Journal of Research in Science Teaching, 32*(2), 195–204.

Fitzhugh, G. (2014). *Techbridge Bay Area program survey findings. 2013–2014 findings*. Lynnwood, WA: Evaluation & Research Associates.

Furger, R. (2002). Building a bridge to science and technology. In M. Chen & S. Armstrong (Eds.), *Edutopia: Success stories for learning in the digital age* (pp. 155–160). San Francisco, CA: Jossey-Bass.

Google. (2014, May). *Women who choose CS—What really matters. The critical role of encouragement and exposure*. Retrieved from http://static.googleusercontent.com/media/www.google.com/en/us/edu/pdf/women-who-choose-what-really.pdf

Harris Interactive for the American Society for Quality. (2009, January 26). *Engineering image problem could fuel shortage: ASQ survey: Career not on radar for kids or parents*. Retrieved from http://www.businesswire.com/news/home/20090126005141/en/Engineering-Image-Problem-Fuel-Shortage#.VgJxWOsaAtw

Kekelis, L., Ancheta, R., & Heber, E. (2005). Hurdles in the pipeline: Girls and technology careers. *Frontiers, 26*(1), 99–109.

Kekelis, L. S., Ancheta, R. W., Heber, E., & Countryman, J. (2005). Bridging differences: How social relationships and racial diversity matter in a girls' technology program. *Journal of Women and Minorities in Science and Engineering, 11*(3), 231–246.

Kekelis, L., & Joyce, J. (2014). How role models can make the difference for girls. *Society of Women Engineers Magazine*, 32–36.

Kekelis, L., & Ryoo, J. (2015, April 22). The other F word: Making sense of failure and nurturing resilience. *Corwin Connect*. Retrieved from http://corwin

-connect.com/2015/04/the-other-f-word-making-sense-of-failure-and-nurturing-resilience/.

Maltese, A. V., & Tai, R. H. (2010). Eyeballs in the fridge: Sources of early interest in science. *International Journal of Science Education, 32*(5), 669–688.

Modi, K., Schoenberg, J., & Salmond, K. (2012). *Generation STEM: What girls say about science, technology, engineering, and math.* New York, NY: Girl Scouts of the USA.

National Center for Science and Engineering Statistics. (2013). *Women, minorities, and persons with disabilities in science and engineering: 2011* (Special Report NSF 13-304). Arlington, VA: National Science Foundation. Retrieved from http://www.nsf.gov/statistics/wmpd/2013/digest/theme4.cfm

Office for Civil Rights. (2012). *Gender equity in education.* Washington, DC: U.S. Department of Education. Retrieved from http://www2.ed.gov/about/offices/list/ocr/docs/gender-equity-in-education.pdf

Pena, M., Kekelis, L., Anaya, M., & Joyce, J. (2013). *Science: It's a family affair. A guide for parents.* Oakland, CA: Techbridge.

Regalla, L. (2010). *SciGirls seven. How to engage girls in STEM.* Retrieved from http://d43fweuh3sg51.cloudfront.net/media/media_files/scigirls_Seven_2013hi.pdf

Rosenberg, H., Wilkes, S., & Harris, E. (2014). Bringing families into out-of-school-time learning. *The Journal of Expanded Learning Opportunities, 1*(1), 18–23.

Stanton-Salazar, R. D. (2011). A social capital framework for the study of institutional agents and their role in the empowerment of low-status students and youth. *Youth & Society, 43*(3), 1066–1109.

STEMconnector and My College Options. (2012). *Where are the STEM students? What are their career interests? Where are the STEM jobs?* Retrieved from http://www.discoveryeducation.com/feeds/www/media/images/stem-academy/Why_STEM_Students_STEM_Jobs_Full_Report.pdf

Techbridge. (2013). *Engage, connect, inspire. How SWE role models can change the world.* Retrieved from www.abdi-ecommerce10.com/swe/c-5-brochures.aspx

The After-School Corporation. (2013). *The 6,000 hour learning gap.* Retrieved from http://www.expandedschools.org/policy-documents/6000-hour-learning-gap#sthash.5850dIeq.dpbs

Traphangen, K., & Traill, S. (2014). *How cross-sector collaborations are advancing STEM learning.* Los Altos, CA: The Noyce Foundation.

Werner, L., & Denner, J. (2009). Pair programming in middle school: What does it look like? *Journal of Research on Technology in Education, 42*(1), 29–49.

CHAPTER 4

EUREKA!-STEM

Hands-On, Minds-On STEM
for At-Risk Middle School Girls

Angie Hodge, Michael Matthews, and Amelia Tangeman

ABSTRACT

The UNO Eureka!-STEM program, which began in 2012, brings female students from lower-socioeconomic environments to the University of Nebraska Omaha (UNO) for an intensive four-week summer program focusing on science, technology, engineering, and mathematics (STEM) the summer before the girls enter grades 8 and 9. The program is a collaborative effort between Girls Inc. and STEM disciplines across UNO, with continuing participants spending the first two summers on UNO's campus, followed by three summers of internship opportunities with local STEM organizations. At UNO the girls gain experience in STEM topics such as rocketry, biomechanics, and "math-emagic." They also have daily lessons on social/emotional development and physical fitness. This chapter describes the program, including academic topics and related field trips, funding information, and collaboration among the faculty, Girls Inc. staff, students, and community partners. Evaluation of the program shows significantly strong increases in perceived content knowledge and increases in expressed interest in technology and engineering careers.

Out-of-School-Time STEM Programs for Females, pages 81–100
Copyright © 2017 by Information Age Publishing

Women, especially racial/ethnic minorities, are still underrepresented in science, technology, engineering, and mathematics (STEM) disciplines. In the spring of 2012, Girls, Inc. of Omaha approached the University of Nebraska Omaha (UNO) to partner in an effort to improve this situation locally and to gain insights to help others address the problem on a larger scale. Thus, since 2012 UNO has collaborated closely with Girls, Inc. of Omaha to host an annual summer camp for middle school girls involved in the local Girls, Inc. organization and who show promise in the STEM disciplines. Girls, Inc. is a nonprofit organization that inspires all girls to be "strong, smart, and bold" through life-changing programs and experiences that help girls navigate gender, economic, and social barriers.

The Eureka!-STEM summer program, held on UNO's campus, uses inquiry-based learning (IBL) strategies to engage the girls in a hands-on, minds-on approach to learning STEM concepts. The girls learn to pose questions about phenomena, develop experiments to test questions, collect data, ask questions about the data, and arrive at conclusions. The Girls, Inc. leadership team, which is composed of Girls, Inc. personnel and UNO faculty (including STEM disciplinary faculty and STEM education faculty and students), work with the girls on a daily basis. Faculty members and students carefully integrate the inquiry-based learning strategies as they create STEM lessons, prepare materials for the camp, and implement the STEM lessons. In addition, Girls, Inc. personnel and UNO faculty members collaborate closely to provide culturally responsive training to prepare student helpers for assisting the faculty members in working with a diverse group of middle school girls. During the camp, the extensive UNO team works together to help teach lessons, assist with field trips, and build relationships with the girls.

This chapter presents synthesized professional literature on females and STEM, selected information on program effectiveness based on data collected about the program, and practical implications that give useful insights into the program. Details are provided on the rationale for the summer program, a description of the program that can aid replication, and data on program successes and challenges.

LITERATURE REVIEW

The need to expose girls, especially those from underrepresented groups, to engaging STEM experiences and to help them see STEM as a real and tangible career path continues to be pressing. Despite knowledge of women's underrepresentation in STEM, growth in STEM fields for women has been relatively stagnant for the past 40 years, even while growth appears in many non-STEM fields (Lippa, Preston, & Penner, 2014). Pryor, Hurtado,

DeAngelo, Blake, and Tran (2009) report that only 17.3% of first-year college women plan on pursuing a STEM major, compared to 32.2% of men. Recently, there is some marginally good news. Meyer, Cimpian, and Leslie (2015) found that gaps between women and men in STEM employment have narrowed in some fields. For example, in molecular biology 54.4% of the 2011 PhD's were female. However, Meyer, Cimpian, and Leslie found that the historical gaps still persist in STEM fields that are considered to require "raw intellectual talent" (i.e., not closely tied to persistence and effort) by people outside and inside the field. For example, the authors found engineering to be such a field based on their survey results and the fact that only 22.2% of the 2011 PhD's were female.

Interventions aimed at the middle school level for girls from underrepresented racial/ethnic minority groups are particularly important. Kinzie (2007) demonstrated that deep gaps in projected STEM career paths appear by at least eighth grade, showing the need for early intervention. Based on their analysis of more than nine years of data across multiple grade levels in Indiana, Beekman and Ober (2015) conclude that making STEM careers a viable option for females from underrepresented racial/ethnic minority groups is particularly vital in the middle grades.

Neither the reasons for the gaps nor what should be done to close the gaps are completely clear. Some fault might lie in the structure of traditional school instruction. For example, Barker and Garver-Doxas (2004) argue that many STEM courses rely on competition and rankings among students rather than collaborative learning, which women tend to prefer. This competitive focus could lead to disengagement and discouragement for women. Further, several research reports indicate that negativity towards STEM held by adults who interact with female students and underrepresented racial/ethnic minorities can transfer to students. For example, Grossman and Porche (2014) found that African American and Latino students perceive their science teachers as not being as supportive of their success as other racial groups. They report an increased rate of "microaggressions" on minority students' STEM competence by their teachers, friends, and family. According to Sue and Constantine (2007), "Racial microagressions are brief, everyday exchanges that send denigrating messages to people of color because they belong to a racial minority group" (p. 137). An example of a microinsult (a form of microaggression) experienced by a student in Grossman and Porche's research is, "Oh, you're Spanish and you're doing math that good?" (p. 710). In terms of workplace perceptions, Chen and Moons (2015) report research indicating that female students believe they will have less power to influence others if involved in a STEM field. Finally, in a major report on how to close STEM employment gaps, the Organisation for Economic Co-operation and Development (2015) concludes:

> As the evidence in the report makes clear, gender disparities in performance do not stem from innate differences in aptitude, but rather from students' attitudes towards learning and their behaviour in school, from how they choose to spend their leisure time, and from the confidence they have—or do not have—in their own abilities as students. (p. 3)

Hence, improving middle school girls' attitudes is an important goal for intervention efforts.

Intervention programs aimed at increasing the number of women in STEM fields have been and continue to be planned, piloted, and refined. Many show promise. However, even seemingly well-crafted intervention programs have shown mixed results.

Some research results show the difficulty of developing interventions that influence participants in intended ways. For example, Archer, DeWitt, and Dillon (2014) report findings from a six-week STEM camp intended to expose middle school female students to STEM careers. They found that the girls' interest in choosing a STEM career had not changed, but their awareness of the possibilities that exist in STEM careers had increased dramatically. One study showed that utility-valued interventions with parents were highly effective in influencing the choices of STEM classes for low-achieving boys and high-achieving girls, but not for low-achieving girls (Rozek, Hyde, Svoboda, Hulleman, & Harackiewicz, 2015). Bystydzienski, Eisenhart, and Bruning (2015) found that a three-year intervention program was not effective in transforming young female minorities' interest in pursuing STEM degrees, despite strong increased interest in STEM fields over the course of the program. Their findings indicate that fear of failure and financial concerns were hampering factors.

However, some promising results also appear in the literature. For example, Stoeger, Duan, Schirner, Greindl, and Ziegler (2013) found that online mentoring successfully helped females ages 11–18 change their short-term and long-term STEM goals as compared to a control group. Also, Dasgupta, Scircle, and Hunsinger (2015) found that when first-year female STEM undergraduates worked in small groups comprised mostly of other females, these microenvironments increased the female students' participation and motivation, as well as influenced their career aspirations toward staying and progressing in STEM fields.

Some global summaries of related research seem promising. Tsui (2007) concludes that successful programs need holistic approaches that employ a variety of strategies. Shapiro and Sax (2011) identify some major factors that contribute to STEM success for women, including (a) family influences and support, (b) interest and sense of belonging in STEM fields, (c) self-efficacy in STEM domains, (d) early exposure to preparatory classes in middle grades and (e) employing teaching strategies that have been shown to support female learning STEM, like IBL, that encourage lots of interaction with peers and teachers and less emphasis on competition.

The Eureka!-STEM program created by Girls, Inc. of Omaha and the University of Nebraska Omaha seeks to take such a holistic approach to encourage underrepresented females' participation in STEM fields. By providing a variety of hands-on, engaging STEM activities at an early stage in girls' lives, Eureka!-STEM strives to address Shapiro's and Sax's (2011) factors to contribute to the success of females in STEM. In particular, Eureka!-STEM intentionally targets culturally relevant teaching, early exposure to STEM careers, interest in STEM subjects, and self-efficacy in STEM domains.

EUREKA!-STEM PROGRAM DESCRIPTION

Knowing the importance of out-of-school-time programs in the success of middle school girls, Girls Inc. of Omaha asked the University of Nebraska Omaha (UNO) to partner with them to establish a five-year Eureka!-STEM program. The University's role in the partnership is to host a four-week summer program on campus during the first two years of Eureka!-STEM. Eureka! is a nationwide program created by Girls, Inc. intended to foster girls' skills in science, technology, engineering, and mathematics while providing opportunities for personal development and physical activity. Eureka! girls begin the program as incoming eighth graders and persist in the program for five years, coming back to campus for a second summer as incoming high school freshman. Girls Inc. staff supervises and leads the last three years of the program. In these final years the girls apply what they have learned on campus by participating in internships in the community. The program is fashioned on a purposeful design to expose girls to STEM concepts and careers, as well as to create a well-rounded citizen, girls who are "strong, smart, and bold," as the Girls, Inc. motto heralds.

The need for a program such as Eureka! is paramount. There is a national shortage of female representation in STEM majors and careers. This is not due to ability or aptitude. Rather, many girls lack interest and confidence in STEM. Eureka!-STEM encourages young women to pursue STEM by giving them opportunities to explore STEM topics early in the program. Later in the program, the young women are matched to appropriate STEM internships to allow the girls to see themselves in those STEM careers. Because the young women first attend the two summer camps on campus as a cohort, they learn within their peer group and are uninhibited to explore and investigate STEM. The girls are surrounded by program staff and partners who support their passion for STEM, believe they can succeed, and challenge them to pursue their STEM interests beyond their five years in Eureka!-STEM.

Program Overview

The cohort for the first Eureka!-STEM program entered its fifth year in 2016. The program began in 2012 with the first group of 30 incoming eighth graders, who are referred to as "Rookies" within the program. UNO and Girls, Inc. collaborated on a five-year grant to fund the program. The principal investigators consist of a team of education and STEM professors. Following the inaugural year, the first group of Rookies returned for their second year in the program, then considered "Vets," and a new class of Rookies began their five-year journey.

Approximately 30 Rookies are recruited each year from the two Omaha Area Girls, Inc. locations to participate in the program. The majority of participants in both the Rookie and Vet classes each year are from under-represented racial/ethnic groups, mainly African American followed by Hispanic. The girls are selected after an interview process held at Girls, Inc. by the Eureka! Coordinator. The girls chosen for the program display passion for STEM and willingness to try new things, and they must have exhibited exemplary behavior throughout their time with Girls, Inc. The girls must receive parental permission to participate in the program and make an effort to commit to the five years that Eureka! requires. Each Rookie class starts with approximately 30 girls, the majority of whom continue in the program. On average only three to five girls do not go into their second year of the program (as Vets). This is often because the girls move away from Omaha, decide that Eureka! is not right for them, or become highly involved in high school sports. However, with the majority returning for their Vet year, program appeal is apparent.

Participants do not pay a program fee to be involved in Eureka! In order to be a member of Girls, Inc., annual dues are required by the national organization, but they are minimal and assistance is offered to families who cannot cover the dues. Funding for the Eureka! program is covered through the grant and UNO was awarded $79,000 for the 2015 summer program by Omaha Girls, Inc. The money from the grant covers the cost of materials for the STEM sessions, compensation for instructors (of mixed gender), salary for UNO student workers (of mixed gender) who assist faculty in the program, use of the swimming pool, rock climbing wall, and kayaks in the Health and Physical Education and Recreation (HPER) building on campus, field trip admission fees, meal cards for participants' daily lunches, and other administrative costs.

Each STEM session of the four-week Eureka! program is differentiated for the Rookies and the Vets. The Rookies receive introductory instruction in STEM areas that are interdisciplinary and hands-on, while the Vets receive advanced instruction in those areas and are exposed to additional enrichment experiences they did not have as Rookies. Major academic topics for the Rookies include robotics, rocketry, "mathemagic," chemistry, aqua-ponics, high altitude ballooning, financial literacy, life science, and others. The Vets continue these topics, as well as tour the Biomechanics Research Building (BRB) on campus and participate in new sessions related to ge-ology, wearable technologies, and computer programming. The Vets also complete a "ME" project, highlighting their time in Eureka! and their goals for the future. Faculty members from across campus create and teach most of the sessions, with a few sessions being taught by a former staff member who currently teaches at a local elementary school.

The Eureka!-STEM program is a collaborative endeavor. Departments across campus synergize to make the program effective and give the girls an interdisciplinary experience. Starting in December, faculty and staff in-volved in the program and from Girls, Inc. meet monthly to plan the sum-mer program. Various departments are represented in Eureka!, including Teacher Education, Biology, Mathematics, Chemistry, Geology, and Biome-chanics. Our partnership with Girls, Inc. also makes the program possible, because we could not do a true Eureka! program without their girls. In ad-dition, local teachers from the community assist in facilitating workshops, such as technology teachers who lead the computer programming and wearable technologies sessions.

The program could also not take place without the student workers and Girls, Inc. staff who assist the facilitators and build relationships with the girls. Undergraduate student workers are employed to work Eureka! exclusively for the month of June. It is a full-time position, paid through the grant, and student workers represent a variety of majors in STEM and education. In addition, dedicated staff from Girls, Inc. work with the girls through Eureka! and provide institutional knowledge and previous bonds with the girls that enrich the program. Volunteers had not been part of the program until the summer of 2015. These student volunteers were part of a Noyce Fellowship of mathematics students who want to be teachers, as well as from the campus organization NE STEM 4 U.

Many instructional sessions are based on modifications of undergradu-ate coursework. For example, most of the chemistry labs that the girls do are ones that college freshman complete in their undergraduate program of study. The girls love to know this and feel special that they are doing the same things as college students! Being in actual labs, participating in col-lege life, such as eating lunch in the Milo Bail Student Center (MBSC), and using campus recreation facilities allow the girls to feel that they belong on campus and help them consider college a real, attainable goal. They cap off their experience each summer by inviting their parents to the Eureka! graduation ceremony where some of their summer accomplishments are highlighted.

As part of the national Girls, Inc. expectations for implementation of an ideal Eureka! program, the girls also participate in social and emotional development activities in the Personal Development hour every day of the camp. Topics include the dangers of smoking, self-defense tactics, sexual health and well-being, healthy relationship building, bike safety, and many others. These sessions, as well as physical activities the girls participate in every afternoon, contribute to these young women's development as a whole, including well-being and physical fitness. Physical activities in HPER include swimming lessons, Zumba, rock wall climbing, wheelchair basketball, kayaking, group fitness, and much more. Every Friday, the Rookies and the Vets go on separate field trips, which align with the content they have learned throughout the week. The Rookies visit Omaha's Henry Doorly Zoo, the Edgerton Explorit Center, the Great Plains Black History Museum, and City Sprouts. Vet field trips include Glacier Creek Preserve, Adventureland, Acieta, Inc., and Blooms Organics.

Each June, Eureka!-STEM participants arrive on the UNO campus Monday through Friday at 8:45 a.m. for four weeks. Every day, excluding Friday field trip days, is typically structured as follows:

9:00–10:00:	Personal Development Hour (all students together)
10:00–11:30:	STEM Session #1 (Rookies and Vets split for respective sessions)
11:30–12:30:	Lunch (all students together)
12:35–2:00:	STEM Session #2 (Rookies and Vets split for respective sessions)
2:00–3:40:	Physical Fitness/Swimming Session (Rookies and Vets split for respective sessions)
4:00:	Dismissal back to Girls, Inc.

Throughout the four-week program many campus facilities are used to create a more extensive college experience for participants. Many sessions are held in the Roskens Hall College of Education's STEM lab, with other sessions held in the Durham Science Center, Biomechanics Research Building, Health and Physical Education and Recreation building, and the Milo Bail Student Center. Girls in the Eureka! program have access to many tools and equipment that faculty use in their own research, such as chemistry lab equipment, virtual reality lab equipment in Biomechanics, prepared animal and plant tissue slides, computers and robots for programming, and video recording devices, to name a few. Materials ordered through the grant include consumable items and items needed to conduct hands-on activities. For example, in order to lead an effective Wearable Technologies session for the Vets, instructors asked for items such as conductive thread, LilyTiny microcontrollers, coin cell batteries, LEDs, and other items for the girls to sew a monster with light-up eyes. This activity teaches electricity and circuitry concepts, while allowing the girls to produce something to take home.

In order to publicize and showcase Eureka! activities, press releases are created through UNO's Marketing and Communications office. They are highlighted on the UNO website and are often picked up by media outlets. The high altitude balloon launch and the mathemagic sessions were both covered by a local news station. The presence of media on campus is always an exciting event and helps to shine light on the important work taking place in the program.

Despite several years of experience running the Eureka! program, it is not without challenges. While funding has not been an issue because of the grant, a plan must be established for when the grant concludes to ensure that Eureka! on UNO's campus is sustainable. Smaller difficulties include typical logistical issues, such as finding classroom space for sessions, arranging field trips and buses, identifying needed materials to order in time for use, and other matters that must be resolved before the June camp begins. On occasion, facilitators run into behavior difficulties with some of these young teenage girls. However, only one student in the previous three years was not allowed to continue in the program. Girls, Inc. has a behavior management system that is carefully observed, which helps to ensure that everyone enjoys the program and has an opportunity to learn.

Once the girls have "graduated" from our summer program, they are placed in internships in the community. Internship placements involve local STEM businesses and organizations, including Union Pacific Railroad, Omaha's Henry Doorly Zoo & Aquarium, the Nebraska Humane Society, the Strategic Air and Space Museum, and many others. While on campus for the first two years of Eureka!, the girls participate in hands-on, minds-on activities covering STEM content in authentic learning environments, including university chemistry and biomechanics labs, while receiving instruction from University faculty who are experts in their content areas. In the future we plan to collect more data on the success of this follow-up to our summer program.

PROGRAM EFFECTIVENESS

A critical component for determining the effectiveness of the UNO Eureka!-STEM program is program evaluation. This component influences program funding and helps us share with others what has worked well and what could be improved in relation to efforts to influence young women in STEM. Thus, we collect data that help us assess program effectiveness. Results of the survey instrument detailed in this chapter provide information on the girls' attitudes toward STEM, perceptions of STEM content knowledge, future plans, and most and least liked aspects of the program.

Participant Information

A pre- and post-survey for the June 2014 Eureka!-STEM program was administered to 18 rising eighth graders and 18 rising ninth graders who participated in Eureka!-STEM for the entire month. The Eureka!-STEM participants' names were coded Rookie (R) for incoming eighth graders or Vet (V) for incoming ninth graders, and each was given a random one- or two-digit number as an identifier.

Research Design

The surveys are designed to provide insight into several areas: (a) the girls' attitudes toward STEM; (b) the girls' perceptions of the degree to which they gained STEM content knowledge; (c) the girls' plans after high school, including their interest in pursuing STEM careers; and (d) the girls' favorite and least favorite parts of the Eureka!-STEM program. The pre-test was administered the morning of the first day of the Eureka!-STEM camp. The post-test was given on the last day of the camp, immediately before the celebratory graduation ceremony. The girls were required to complete the survey as part of the program expectations, but as some girls were not present the first or last day of camp, a 68% response rate for girls who completed both pre- and post-test was attained.

Instrumentation

Participants completed pre- and post-test surveys in which they rated statements on a five-point Likert scale ranging from 1 (Strongly Disagree) to 5 (Strongly Agree) and answered several open-ended questions. They used iPads displaying the link to the survey to complete the instrument. The surveys, which may be found in the Appendix, were created using SurveyMonkey software.

Data Analysis

The mean response for each question was calculated from the raw survey data. The pre- and post-survey means were then compared to look for trends in the data, and two-tailed *t*-tests were performed to determine whether any changes were statistically significant. Responses to open-ended questions were examined for themes.

Findings

Dispositions Toward STEM

Table 4.1 displays mean responses for the Rookies and the Vets from the pre- and post-survey questions regarding attitudes toward STEM. The table also displays increases or decreases in ratings from the first to second survey. All analyses of potentially significant change in mean scores are based on paired-sample, two-tailed *t*-tests.

The data show that the Vets typically start with higher scores in each category than the Rookies, likely due to their previous experience in Eureka!-STEM and perhaps their additional year in school. Both Rookie and Vet scores tend to increase from the beginning to the end of the camp, indicating greater enjoyment of doing STEM activities.

Content Knowledge

Table 4.2 displays mean ratings for Vets and Rookies on the pre- and post-survey items regarding participants' perceptions of how well they know STEM content in the areas of robotics, rockets, chemistry, physics, and life science. The table also shows increases or decreases in perceived STEM content knowledge from the beginning to the end of the camp.

TABLE 4.1 Mean Participant Ratings of Enjoyment Doing STEM Activities at Camp Beginning and End				
Survey Items	I like doing science activities.	I like doing technology activities.	I like doing engineering activities.	I like doing mathematics activities.
Pre-Survey				
Rookies	3.7	3.6	3.5	3.3
Vets	3.9	4.1	3.6	3.4
Post-Survey				
Rookies	3.8 (+0.1)	4.0 (+0.4)	3.6 (+0.1)	3.3 (+0.0)
Vets	4.2 (+0.3)	4.1 (+0.0)	3.8 (+0.2)	3.8 (+0.4)*

Note: Raw scores range from 1 (Strongly Disagree) to 5 (Strongly Agree). Rating changes from pre- to post-test appear in parentheses.

* $p < 0.05$.

TABLE 4.2 Mean Participant Ratings of Perceived Knowledge of STEM Content at Camp Beginning and End

Survey Items	I know about robots.	I know about rockets.	I know about chemistry.	I know about physics.	I know about life science.
Pre-survey					
Rookies	2.4	2.4	2.8	2.8	3.6
Vets	3.2	3.6	3.4	3.2	3.6
Post-survey					
Rookies	3.6 (+1.2)**	3.6 (+1.2)***	3.4 (+0.6)	3.6 (+0.8)*	3.7 (+0.1)
Vets	4.2 (+1.0)***	3.9 (+0.3)*	4.0 (+0.6)	3.9 (+0.7)*	4.2 (+0.6)*

Note: Raw scores range from 1 (Strongly Disagree) to 5 (Strongly Agree). Rating changes from pre- to post-test appear in parentheses.
* $p < 0.05$, ** $p < 0.01$, *** $p < 0.001$.

The survey results show that participants tended to perceive increases in their STEM content knowledge from program beginning to end. The largest increases from pre-test to post-test appeared in robotics and rocketry content for both groups, and in robotics for the Vets. These figures indicate that the program's attention to these areas, intensive during the first two weeks, appears to be effective.

Career plans. Table 4.3 displays mean ratings for the Rookies and the Vets from the pre-and post-survey questions regarding their interest in STEM

TABLE 4.3 Mean Participant Ratings of Interest in STEM Careers at Camp Beginning and End

Survey Items	I might want a career in science.	I might want a career in technology.	I might want a career in engineering.	I might want a career in mathematics.
Pre-survey				
Rookies	2.9	3.1	2.4	2.9
Vets	3.4	3.1	3.1	3.2
Post-survey				
Rookies	3.1 (+0.2)	3.3 (+0.2)	3.1 (+0.7)*	3.1 (+0.2)
Vets	3.9 (+0.5)	3.8 (+0.7)**	3.3 (+0.2)	3.3 (+0.1)

Note: Raw scores range from 1 (Strongly Disagree) to 5 (Strongly Agree). Rating changes from pre- to post-test appear in parentheses.
* $p < 0.05$, ** $p < 0.01$.

careers. The table also shows increases or decreases in expressed interest from the first to the second survey.

The data show that both the Rookies' and the Vets' interest in STEM careers increased from the beginning to the end of Eureka!-STEM. The Rookies' largest increase was in engineering, while the Vets' largest increase was in technology. It is interesting to note the change in engineering career interest for the Rookies. On the pre-survey, many Rookies indicated that they looked forward least to the engineering sessions of the camp. However, on the post-survey the Rookies responded that one of the content areas they enjoyed most was that of engineering. As engineering is a STEM field that has historically been driven by white males, this is an inspiring outcome of the program. One participant wrote on her post-survey: "I think more girls should become engineers because that's a more man-dominated job." Some of the girls recognize where the need is and how they can fill those needs in the future.

One possible reason for a large gain in engineering interest for the Rookies was that a strong mentor led the engineering sessions. This mentor is a UNO professor of engineering education who is an underrepresented minority. He had worked for several years in the field before teaching middle school youth in a low-SES school. He presently uses his experiences in engineering and teaching underrepresented students similar to that of many Eureka!-STEM girls to provide a learning environment to which the girls respond well. His personal stories about his career in engineering are inspirational, and the girls see this faculty member as someone they can approach to ask advice about careers in engineering.

After the conclusion of the Eureka!-STEM summer program, 67% of the Rookies and 72% of the Vets reported an interest in a STEM career (i.e., rated them 4 or 5) These proportions are inspiring, as they reflect the girls envisioning themselves in STEM careers. With the resources provided through the Eureka!-STEM program, the girls acquire tools to pursue those goals.

Open-response questions.   The post-survey also asked the girls to answer some open-ended questions about their future plans, as well as what they liked most and least about Eureka!-STEM. When asked what they wanted to do after they graduated high school, most participants responded that they want to go to college. Only four stated that they were unsure, two responded that they wanted to go into the military before college, and three said they wanted to get their PhD. When asked what they thought girls might contribute to STEM careers, many girls stated that girls could offer a different perspective. One participant noted that girls can "use different ideas and show that girls are smart too."

When asked what they liked most about the program, the girls tended to say they enjoyed the engineering activities, many specifically mentioning the professor who taught the sessions. (A few mentioned swimming.) When asked what they liked least about the program, many girls responded that they did not like all the walking to and from buildings on campus, or when other girls caused "drama" during the month. For example, one day two girls told another girl they did not want to work with her anymore. For the next hour or so, there was noticeable tension in the class between two groups of girls based on who they sided with, but these issues are minimal. The Eureka!-STEM coordinators handle each situation individually. The girls are given three chances in relation to inappropriate behavior before they are no longer permitted to return to the program. This is a rare occurrence, as the girls are hand picked to be in the Eureka!-STEM program and are among the most well-behaved and motivated at Girls, Inc.

Another survey question asked the girls how well they thought they performed on the STEM content during the camp. Most responses were positive. The girls commented that they felt proud of how they did and that they did a good job. One girl stated, "I think I did really good. I even impressed some people." Another commented, "I thought that STEM was going to be hard but they make it easy to understand."

The survey responses provide valuable participant input that allows the program coordinators and facilitators to continue to improve programming. The girls appear to have taken the survey seriously, as evidenced in their thoughtful answers, and the information gathered points to the effectiveness of the UNO Eureka!-STEM program.

KEY PROGRAM ELEMENTS

Although the Eureka!-STEM program currently runs smoothly and is highly organized, it took time and trial and error to get to this point. The following practical implications and suggestions provide insight into running a STEM camp for young women based on our experiences conducting the program for five years.

Designate a Coordinator

When a project involves collaboration with several key players, many coordination, communication, and logistical issues must be handled to ensure a successful camp. We recommend that a summer program such as Eureka!-STEM have a dedicated coordinator who is the liaison between the university and the partnering organization, as well as the person responsible for scheduling sessions, training student workers, organizing and presiding over planning meetings, and handling daily camp logistics. This is by no means an all-inclusive list, but this person plays a critical role in that she/he is the point of contact and is involved with the camp from the first planning meeting to the last day of camp. The UNO Eureka!-STEM Coordinator works closely with the Eureka-STEM! Coordinator for the partnering organization (Girls, Inc. of Omaha) and is a large part of all decision making related to the camp.

The Eureka!-STEM camp did not have a full-time staff member identified as coordinator for the first two years. While the leadership team was able to band together to conduct successful camps prior to designation of a program coordinator, the camps have run much more smoothly with one person who focuses on all the tasks associated with such a large endeavor as a month-long summer camp.

Provide Experiences for Future Teachers

We wrote and were awarded a National Science Foundation Noyce grant that included scholarships for potential pre-service mathematics teachers who are college freshmen or sophomores to gain direct experience teaching young people. In serving that goal and seeking to improve Eureka!-STEM, we provide our Noyce scholars an opportunity to work with the Eureka!-STEM girls, who in turn get to learn from young mentors to whom they can better relate.

As part of the Omaha Noyce Partnership, six Noyce interns per year now assist the Eureka!-STEM leadership team, which is composed of Girls, Inc. personnel and UNO faculty, including STEM disciplinary faculty and STEM education faculty. Each intern works 40 hours per week for six weeks, including one week of training, the four weeks of the camp, and one week of follow-up activities. During training, interns are paired with a faculty mentor and introduced to IBL strategies as they help create STEM lessons and prepare materials for the camp. Girls, Inc. personnel and UNO faculty also collaborate to provide culturally responsive training to prepare the interns for working with a diverse group of middle school girls. During the camp, interns teach lessons, assist on field trips, and gain experience working with underrepresented students.

Build Partnerships

A foundational element of the Eureka!-STEM program is providing opportunities for the girls to learn about STEM careers through role models and by visiting local STEM organizations and businesses. The partnership among departments across UNO and with Girls, Inc. is crucial to the program's success, and so are partnerships forged with STEM representatives and organizations within the Omaha community. Eureka!-STEM relies on these community connections to provide speakers to conduct presentations for the girls during many of the personal development sessions or to host the girls for STEM field trips.

It is much more meaningful for the girls to hear personal stories or visit institutions firsthand, instead of just hearing about available STEM opportunities. The girls are often able to network with the STEM individuals who serve as guest speakers; they may ask questions and receive contact information to continue conversations that begin at the camp. Many of these partnerships have led to new internship positions being offered to the girls in their third through fifth years in Eureka!-STEM. Thus, making connections and creating partnerships can greatly enhance the effectiveness of a program such as ours.

CLOSING COMMENTS

As described in this chapter, the University of Nebraska Omaha has collaborated closely with Girls, Inc. to host an annual summer camp for middle school girls involved in Girls, Inc. and who show promise in STEM. Girls, Inc. is a nonprofit organization that inspires all girls to be "strong, smart, and bold" through life-changing programs and experiences that help girls navigate gender, economic, and social barriers. In 2011, Girls, Inc. served almost 1,500 girls (many from the Omaha Public Schools) at its two Omaha centers and through its outreach programs. Of these, more than 90% were from underrepresented racial/ethnic groups and more than 70% came from families whose annual household income was less than $30,000. In the Eureka!-STEM summer camp held at UNO, inquiry-based learning strategies are used to engage the girls in a hands-on approach to learning STEM concepts.

The collaborative environment not only exposes young women to STEM at an early age, but it engages them in STEM beyond the standard school day. These young women have a chance to do STEM, to learn about STEM careers, and to learn how to make decisions in school and life that will help them be successful in a STEM career.

Program evaluation data show that Eureka!-STEM might play an important role in increasing women in STEM. For one thing, the program might help camp participants visualize themselves as part of a college campus. The large percentage (49 of 53, or 92%) who say they plan to attend college is encouraging and contrary to trends reported in the literature. Further, the Eureka!-STEM program seems to provide positive experiences in STEM settings. The young women report that they like the program and indicate increased interest in STEM fields. It is possible that their positive interactions with the university STEM faculty and students counteract the "microassaults" they might experience in other education settings (see Grossman & Porche, 2014). Finally, girls who participate in this program show increased interest in STEM careers by the end of the program, with approximately 70% of the girls indicating interest in pursuing a STEM field in college. While Eureka!-STEM is only the introductory component of a larger multi-year strategy implemented by Girls, Inc., changes in the future plans of the young women in the program already appear, especially in relation to engineering and computer science. These results are especially important, as these two fields still experience major underrepresentation of females in STEM compared to many biology fields (Meyer, Cimpian, & Leslie, 2015).

Longitudinal data to be collected in years to come will show the true influence of the program. We hope to follow the girls and find out if our camp helped the girls in their academic careers. We will track (a) high school graduation rates, (b) university attendance, (c) choice of college major, and (d) college degree persistence. Ideally, follow-up investigations will include both surveys and focus groups to better understand participant "stories."

In the meantime, our results from survey data are promising. Our program participants understand that women are underrepresented in STEM fields and that programs such as ours provide experiences that might help close those gaps. The girls enjoy the program and return year after year. The fact that interest in the program continues to grow, with enrollment becoming more competitive, shows we are making steps in the right direction.

Through careful planning and ongoing improvements, the UNO Eureka!-STEM program will continue to improve each year. We are pleased to serve as a model for others and hope this book chapter will serve as a catalyst for others to start similar programs in their communities. We, the authors, are happy to answer any questions about our program via email.

APPENDIX

Pre- and Post-Eureka!-STEM Survey Questions

Open Answer: What is your name?

1. Open Answer: What is your age?
2. Please check: Are you a Rookie or a Vet?
3. Likert Scale: Using a scale of 1–5, with 1 being Strongly Disagree and 5 being Strongly Agree, answer the following questions: I like doing science activities, I like doing technology activities, I like doing engineering activities, and I like doing mathematics activities.
4. Likert Scale: Using a scale of 1–5, with 1 being Strongly Disagree and 5 being Strongly Agree, answer the following questions: I know about robots and how they work in real life, I know about rockets and how they work in real life, I know about chemistry and what it is in real life, I know about physics and what it is in real life, and I know about life science and what it is in real life.
5. Open Answer: What do you want to do when you graduate high school?
6. Open Answer: What do you think that girls in particular can do in STEM?
7. Open Answer Pre-Test: What are you looking forward to most in Eureka!?
8. Open Answer Post-Test: What did you enjoy most in Eureka!?
9. Open Answer Pre-Test: What are you looking forward to least in Eureka!?
10. Open Answer Post-Test: What did you enjoy least in Eureka!?
11. Likert Scale: Using a scale of 1–5, with 1 being Strongly Disagree and 5 being Strongly Agree, answer the following questions: I might want a career in science one day, I might want a career in technology one day, I might want a career in engineering one day, and I might want a career in mathematics one day.

REFERENCES

Archer, L., DeWitt, J., & Dillon, J. (2014). 'It didn't really change my opinion': Exploring what works, what doesn't and why in a school science, technology, engineering and mathematics careers intervention. *Research in Science & Technological Education, 32*(1), 35–55. doi:10.1080/02635143.2013.865601

Barker, L. J., & Garvin-Doxas, K. (2004). Making visible the behaviors that influence learning environment: A qualitative exploration of computer science classrooms. *Computer Science Education, 14*(1), 119–145.

Beekman, J. A., & Ober, D. (2015). Gender gap trends on mathematics exams position girls and young women for STEM careers. *School Science & Mathematics, 115*(1), 35–50. doi:10.1111/ssm.12098

Bystydzienski, J. M., Eisenhart, M., & Bruning, M. (2015). High school is not too late: Developing girls' interest and engagement in engineering careers. *Career Development Quarterly, 63*(1), 88–95. doi:10.1002/j.2161-0045.2015.00097.x

Chen, J. M., & Moons, W. G. (2015). They won't listen to me: Anticipated power and women's disinterest in male-dominated domains. *Group Processes & Intergroup Relations, 18*(1), 116–128. doi:10.1177/1368430214550340

Dasgupta, N., Scircle, M. M., & Hunsinger, M. (2015). Female peers in small work groups enhance women's motivation, verbal participation, and career aspirations in engineering. *Proceedings of the National Academy of Sciences, 112*(16), 4988–4993. Retrieved from http://www.pnas.org/content/early/2015/04/03/1422822112

Grossman, J. M., & Porche, M. V. (2014). Perceived gender and racial/ethnic barriers to STEM success. *Urban Education, 49*(6), 698–727. doi:10.1177/0042085913481364

Kinzie, J. (2007). Women's paths in science: A critical feminist analysis. *New Directions for Institutional Research, 133,* 81–93.

Lippa, R. A., Preston, K., & Penner, J. (2014). Women's representation in 60 occupations from 1972 to 2010: More women in high-status jobs, few women in things-oriented jobs. *Plos ONE, 9*(5), 1–8. doi:10.1371/journal.pone.0095960

Meyer, M., Cimpian, A., & Leslie, S. (2015). Women are underrepresented in fields where success is believed to require brilliance. *Frontiers in Psychology, 6,* 1–12. doi:10.3389/fpsyg.2015.00235

Organisation for Economic Co-operation and Development (OECD). (2015). *PISA 2012 results: The ABC of gender equality in education: Aptitude, behaviour, confidence.* Paris: OECD. Retrieved from http://www.oecd.org/pisa/keyfindings/pisa-2012-results-gender.htm

Pryor, J. H., Hurtado, S., DeAngelo, L., Blake, L. P., & Tran, S. (2009). *The American freshman: National norms fall 2009* [Expanded ed.]. Los Angeles: University of California, Los Angeles, Higher Education Research Institute.

Rozek, C. S., Hyde, J. S., Svoboda, R. C., Hulleman, C. S., & Harackiewicz, J. M. (2015). Gender differences in the effects of a utility-value intervention to help parents motivate adolescents in mathematics and science. *Journal of Educational Psychology, 107*(1), 195–206. doi:10.1037/a0036981

Shapiro, C. A., & Sax, L. J. (2011). Major selection and persistence for women in STEM. *New Directions for Institutional Research, 152,* 199–145.

Stoeger, H., Duan, X., Schirner, S., Greindl, T., & Ziegler, A. (2013). The effectiveness of a one-year online mentoring program for girls in STEM. *Computers & Education, 69,* 408–418. doi:10.1016/j.compedu.2013.07.032

Sue, D. W., & Constantine, M. G. (2007). Racial microaggressions as instigators of difficult dialogues on race: Implications for student affairs educators and students. *The College Student Affairs Journal, 26*(2), 136–143.

Tsui, L. (2007). Effective strategies to increase diversity in STEM fields: A review of the research literature. *The Journal of Negro Education, 76*(4), 555–581. Retrieved from http://www.jstor.org/stable/40037228

CHAPTER 5

OUT-OF-SCHOOL-TIME STEM SUPPORT FOR GIRLS IN THE SILVER STATE

Lynda R. Wiest and Heather Glynn Crawford-Ferre

ABSTRACT

In this chapter, we describe a one-week residential summer camp for North-ern Nevada rising seventh- and eighth-grade girls. The camp, which focuses on mathematics and technology, serves girls of varying abilities and from diverse backgrounds. Evaluation data reported include results of a pre- and post survey consisting of ratings involving beliefs about mathematics as a gen-dered domain and open-ended questions regarding key characteristics of the program itself (completed by 31 and 99 girls, respectively). Participants showed a significant favorable change in their perception of mathematics as a female domain from the camp beginning to end. In terms of salient pro-gram components, the girls named academic aspects of the camp as the most important program component and secondarily, but to a much lesser de-gree, the social and recreational aspects of the camp. The meaning of these findings and the potential value of an out-of-school-time STEM program for females are discussed.

Out-of-School-Time STEM Programs for Females, pages 101–121
Copyright © 2017 by Information Age Publishing

The United States lacks a well-prepared workforce in science, technology, engineering, and mathematics (STEM) careers (Heaverlo, 2011; Paulsen, 2013). Workers in these fields tend to be White and Asian men (McGee, 2013). A greater number of females, individuals from underrepresented racial/ethnic groups and low-income backgrounds, and people with disabilities are needed to participate in and contribute to STEM occupations (Davis & Hardin, 2013; White, 2013). However, U.S. students show weak STEM performance in international contexts (Davis & Hardin, 2013; Dillivan & Dillivan, 2014), and underrepresented groups face additional challenges to STEM success. Females, for example, grapple with less favorable dispositions toward STEM than males, including weaker self-beliefs and confidence, as well as lower interest (Goetz, Bieg, Lüdtke, Pekrun, & Hall, 2013; Lubienski, Robinson, Crane, & Ganley, 2013; Ross, Scott, & Bruce, 2012). Indeed, it appears that females' dispositions contribute more to the gender gap in STEM than achievement, although dispositions can influence STEM achievement and participation (Goetz et al., 2013; Riegle-Crumb, King, Grodsky, & Muller, 2012; Ross et al., 2012). A survey conducted with 1,200 middle school boys and girls showed that more than twice as many boys as girls chose STEM for future careers, and the same 2-to-1 ratio of boys to girls said their parents would support such a choice (Shapiro, Grossman, Carter, Martin, Deyton, & Hammer, 2015). Actual participation rates tend to mirror expressed intent. For example, high school students in a selective one-week physics program were 74% male and 68% White (Constan & Spicer, 2015), and low-income college students are about half as likely as higher-income students to remain in their declared STEM majors (Johnson & Walton, 2015).

The need to support underrepresented groups in STEM to fuel the STEM workforce and to provide opportunities for better life quality for individuals from these groups is apparent. In this chapter, we focus on females as one such group, with awareness of the compounding effects of intersecting identities, such as race/ethnicity and socioeconomic status. We discuss the potential contributions of an out-of-school-time (OST) program, specifically, a one-week residential summer camp, as one strategy for supporting females in STEM.

RELATED LITERATURE

Factors That Influence Females in STEM

As noted earlier, females' dispositions toward STEM are particularly problematic as a factor that contributes to lower achievement and participation in STEM. It is therefore important to determine what influences these dispositions. The culprit seems to be sociocultural influences, in other words, the way we have constructed the world to place girls at odds with STEM subjects and with their own capabilities in these disciplines. In the wider societal sense, gendered patterns in STEM might in large part be due to this social structure rather than to individual skills (Riegle-Crumb et al., 2012). Riegle-Crumb et al. (2012) place the blame squarely on "societal pervasiveness of gender essentialist beliefs and the accompanying socialization and micro-level interactions that support them" (p. 1067).

Data from a survey of 1024 high school students and interviews with 53 of those students show that females report experiencing disparaging comments and behaviors that stereotype them and question their competence and belonging in STEM (Grossman & Porche, 2014). The study authors state, "Both female and male participants recounted gender role expectations for girls and women that were antithetical to STEM interest and achievement, such as family members stating that some careers are not appropriate for women" (Grossman & Porche, 2014, p. 717). Teachers' gendered perceptions, too, might contribute to the damaging messages females receive (Lubienski et al., 2013).

STEM Out-of-School-Time Programs for Females

Students have been shown to lose ground in their academic skills over the summer months while out of school, which is particularly true for students from low-income families and in the subjects of mathematics and reading (McCombs, Augustine, Schwartz, Bodilly, McInnis, Lichter, & Cross, 2012; Slates, Alexander, Entwisle, & Olson, 2012). It is thus believed that structured learning opportunities, such as those that take place in school, are important to academic growth and that the summer months can be further used to that advantage by involving youth in out-of-school-time programs (Slates et al., 2012). Accordingly, the summer period can serve to widen or narrow achievement gaps (McCombs et al., 2012; Slates et al., 2012). McCombs et al. (2012) note, "Participation in summer learning programs should mitigate learning loss and could even produce achievement gains. Indeed, educators and policymakers increasingly promote summer learning as a key strategy to improve the achievement of low-performing students" (p. 47).

Out-of-school-time programs can foster more favorable dispositions, such as confidence and interest, as well as enhance disciplinary competence and encourage future participation in STEM among struggling and underrepresented students (Constan & Spicer, 2015; Davis & Hardin, 2013; Heaverlo, 2011; McCombs et al., 2012; Mohr-Schroeder et al., 2014). The middle through high school years might be a particularly important juncture for engaging girls in STEM and building STEM awareness (Constan & Spicer, 2015; Mohr-Schroeder et al., 2014), although upper elementary students are also "ripe" for such attention in that they are beginning to work with more complex academic material (Dillivan & Dillivan, 2014). In the remainder of this chapter, we describe a summer mathematics and technology program for middle-grades girls and selected outcomes identified through program evaluation measures.

PROGRAM DESCRIPTION

The Northern Nevada Girls Math & Technology Program (http://www.unr.edu/girls-math-camp) started in 1998 under Lynda Wiest, who continues to direct the program, with Heather Crawford-Ferre now serving as Assistant Director. The program was developed to address concerns about females' underrepresentation in and potentially unfavorable dispositions toward mathematics and technology. Specifically, the program's purpose is to increase girls' knowledge, skills, and confidence in mathematics and technology in order to enhance mathematical and technological competence in girls' personal, academic, and occupational lives. The program consists of the following three components: (a) a one-week overnight summer camp held at the University of Nevada, Reno; (b) once-a-week after-school computer programming sessions conducted at a local school; and (c) a website that provides information, resources, and a password-protected discussion area.

Program Participants

Northern Nevada girls of all abilities and backgrounds who are rising seventh or eighth graders (i.e., will enter those grades in the fall) are eligible to participate in the summer camp, which to date has been the main program component. The middle-grades level was chosen as an important crossroads at which to support girls in these disciplines because of heightened sensitivity to gender roles during adolescence and the resulting influence on girls' achievement and attitudes. Further, this age is an important time to make decisions about future courses and to begin considering career paths. Application forms are sent to all Northern Nevada public, private, charter, and tribally operated schools that house the relevant grade levels, as well as the local Girls Scouts and Boys & Girls Club. The form is also sent to the Nevada Homeschool Network and posted on the program website.

Sixty girls (30 for each grade level) are chosen to attend the camp. For the older group, girls who attended the previous year have first priority, those who applied but were not chosen the previous year have second priority, and the remaining slots are filled with new applicants. New applicants for the older group and all applicants for the younger group are chosen randomly (using a random number generator) from among applicants who meet the application deadline. Those who do not get accepted initially are prioritized on the wait list, and several girls from each grade level are typically admitted from that list, as a few girls often drop out up until the day the camp starts and sometimes during the camp week itself. Girls who participate in the summer camp may participate in a password-protected Edmodo site (social media site that is embedded in the website and governed by rules of participation that are predominantly academic in nature).

The newest program component, one year old at this writing, is an after-school computer programming club for 20 fourth- and fifth-grade girls who attend a Title I school and participate in an after-school program during that time, which allowed them to easily shift into our computer club. The one-hour class is conducted every Friday in the school's computer lab. The girls learn to use *Terrapin Logo*'s computer programming language to program a robotic car with an inserted writing implement. They create practical products, such as greeting cards, because useful or contextualized STEM applications are known to motivate females (e.g., Ashcraft, Eger, & Friend, 2012; Lum, 2015). Hereafter, this chapter will focus on the long-standing summer camp.

Program Structure

The summer camp is residential because it better allows girls who do not live close to the college campus where the program is held to participate. The girls stay in a residence hall on campus and eat their meals in the college cafeteria. Campers only need transportation to and from the camp Sunday afternoon and Friday afternoon, with an option for program staff to facilitate carpooling. The overnight aspect further allows time for recreational activities and informal interaction with peers and staff in ways that build important social networks and inspire mentoring, connections that support academic efforts. The academic portion of the camp starts at 8:45 a.m. and ends at 4:15 p.m. most days. A brief mid-morning and mid-afternoon break provides a needed mental rest and social outlet with a reasonably healthy snack. (See the sample schedule in Table 5.1.) On three camp days, the girls may choose from structured, supervised spare-time activities of one to one and one-quarter hours each before and after dinner. Options vary but usually include computer lab time, dorm time, and outdoor recreation, and on one occasion a visit to the campus store. Evening recreation includes one night roller skating or swimming at an indoor pool, a movie

TABLE 5.1	Sample Thursday Schedule for Girls Entering Grade 7
8:45 a.m.	Problem solving/reasoning
9:15 a.m.	Guest speaker on her use of math/technology in the workplace
10:00 a.m.	Data analysis and probability
10:30 a.m.	Snack break
10:45 a.m.	Data analysis and probability (continued)
11:45 a.m.	Spatial-reasoning task
12:00 p.m.	Lunch
1:00 p.m.	Using geometry software (*Terrapin Logo*)
2:30 p.m.	Snack break
2:45 p.m.	Geometry
4:15 p.m.	Spare time
5:30 p.m.	Dinner
6:30 p.m.	Spare time
8:00 p.m.	Camper talent show

Note: Printed schedules for program participants include instructor/speaker names for each segment, room locations, and options available for spare-time segments.

night, an afternoon at a water park followed by miniature golf and a go-kart ride, and a talent show.

Instructional Content and Approaches

Content chosen for the program includes topics considered particularly important for students in general (such as content that is especially valuable to their future or that might receive inadequate attention in school) or for girls in particular (e.g., areas in which they tend to demonstrate weak performance and/or dispositions). Both camper groups do geometry/measurement, problem solving, and spatial skills, and both are exposed to female role models in mathematics and computer science. The older girls also learn algebra and use of *Geometer's Sketchpad*, whereas the younger girls have lessons on data analysis/probability and *Terrapin Logo*. (Computer segments take place in the computer lab with one girl to a computer.) Other technologies variably incorporated into lessons based on instructor choice include basic ("four-function") or graphing calculators, tablet and handheld computers, and student response systems. This aspect of the program fits with Newbill, Drape, Schnittka, Baum, and Evans' (2015) recommendation to incorporate technology as one important feature of designing OST curriculum.

Instructors may choose any subset of the Nevada Academic Content Standards for Mathematics within the camp's designated topics for the forthcoming grade level. Augustine and McCombs (2015) consider such alignment with school curriculum to be important for OST programs. We address regular school content but in more engaging and thorough ways than might be practical in schools due to time constraints imposed by hefty school curricula and accountability for strong standardized test performance. We strive for teaching the material meaningfully and in a logically sequenced manner, supported by technological tools. We take an investigative approach toward learning but do not use large, involved projects.

Instructors may teach their content in the way they see fit but are given general guidance regarding instructional approaches known to serve girls well in mathematics. The main approaches involve use of collaborative group work, student communication (e.g., discussion while doing mathematics and explanation of solution methods and answers), technology use, and active, hands-on learning (Ashcraft et al., 2012; Dillivan & Dillivan, 2014; Mohr-Schroeder et al., 2014). The girls in each of the two classrooms work within randomized groups of four that change daily.

Other methods foundational to the camp are use of a female-only environment with exposure to female role models and mentors and some attention to career awareness and preparation (Ashcraft et al., 2012; Mohr-Schroeder et al., 2014; Shapiro et al., 2015). In an all-camp session, a female guest speaker discusses how she uses mathematics and/or technology on the job. We ask her to discuss: the types of work she does for her job, including a specific project as an example; why she chose her career; what she does and does not like about her job; preparation needed to enter her career; any difficulties she encountered en route to or during her career and how she overcame them; and how she balances professional and personal life (while briefly noting her leisure activities). We say it is helpful to include some visual aids (e.g., photos) and to consider involving the girls in doing a brief task or discussing a question at least once. In another all-camp session, the girls explore biographical information about historical and contemporary women in STEM. (Both of these sessions last 45 minutes.)

Staff

Typically, staff members are female. The only exception was a male doctoral student who served as a behind-the-scenes program assistant for one year. Besides the Director and Assistant Director, who do not occupy instructional roles, the main program staff are instructors, who are seasoned teachers that teach the main lessons, and instructional assistants (two per classroom) who are upper-division undergraduates or early-career teachers and who assist instructors during their lessons, supervise the girls around the clock, and conduct the problem-solving and spatial skills segments. Instructors are carefully chosen by researching local teachers who typically have substantial teaching experience, maintain a focus in mathematics education (e.g., have attained or are in the process of attaining a graduate degree in that area and/or play a leadership role in PK–12 mathematics education), and have a reputation for being an effective, student-centered teacher as determined by peers, school district mathematics education leaders, and teacher educators. Instructional assistants are selected on the basis of their strong performance in undergraduate mathematics methods courses. All instructional staff are individuals considered to have strong content knowledge, effective instructional approaches, favorable dispositions toward teaching and mathematics (e.g., hold high expectations for all students), and personalities that suit adolescent girls (e.g., enthusiastic, supportive, and sensitive to diverse girls). Instructors and instructional assistants tend to remain in their positions for several or many years because they enjoy working with and are very dedicated to the program.

Teacher selection and training is considered highly important for out-of-school-time programs (Augustine & McCombs, 2015; Mosatche, Matloff-Nieves, Kekelis, & Lawner, 2013; Newbill et al., 2015). For this program, although instructor selection is very purposeful, instructors are only given guidance in writing as to key instructional practices considered important for girls and this type of program, but they receive no further training. This differs from some recommendations for effective OST programs. The reason we do not provide staff training is because we choose highly effective instructors from the start, we want instructors to have the creative latitude to teach the material in their own way within our general guidelines so they have greater investment in and ownership of their instruction, and we recognize that training staff requires additional resources (human, monetary) that we do not currently possess.

About one to three volunteers work in each of the two classrooms during the week, assisting the main instructors and the program at large in various ways. Volunteers are typically college students or practicing teachers, but they might also be high school students, parents, or other community members. Undergraduate or graduate students sometimes earn independent study credits by doing scholarly reading and writing in conjunction with classroom observations during the camp week.

Parent Involvement

Parents are invited to engage in the opening and closing segments of the summer camp. After registration on opening night (Sunday evening), the Program Director (or Assistant Program Director in her absence) gives a 45-minute welcome address to the girls and their parents. During this session, which takes place in a large auditorium-style room, we introduce the staff, show a slideshow of the previous year's camp, give an overview of the camp (e.g., history, purpose, approach), provide some factual information about females in STEM, make various announcements directed toward parents (e.g., regarding camper supervision, opportunities to visit the camp or communicate with daughters, and ways to support and encourage daughters in STEM), and announce selected guidelines to the girls (e.g., behavioral expectations, including no cell phone use or consumption of energy drinks). We also announce and display the program sponsors.

Immediately after this session, we move to a large open multipurpose room where designated staff or volunteers have prepared light snacks for campers, family members, and staff to enjoy while walking around for an informal "meet and greet." Staff members look for girls who seem to be alone, shy, or distressed during this time to talk with them and to try to connect them with other girls. This is also a time for parents to meet and talk with staff to provide some reassurance for their own potential apprehension about leaving their daughters for the week. The parents are then asked to leave so we may begin some introductory activities with the campers.

Parents may also participate in the closing camp session on Friday afternoon. The instructional assistants are provided time in the early afternoon to help volunteer campers prepare brief overviews of what they learned in the main topic areas for the week. For this final parent session, both camper groups move to a larger room, where the instructional assistants welcome the parents and conduct two final camp components. For the first, which lasts about 20–30 minutes, volunteer individuals, pairs, or small groups demonstrate something they learned in each of the main instructional segments. For the computer class, for example, this might include displaying and explaining an animated figure created in *Geometer's Sketchpad* that is preloaded onto the classroom computer. Including such a "public presentation" is a strategy recommended by Newbill et al. (2015). The second main component of the final parent session involves problem solving where girls team with parents for 20–30 minutes to solve a rich word problem and then share strategies and solutions as a whole group. (Girls whose parents do not attend this session join friends' families or are paired with available staff or volunteers.)

In the past, we provided snacks at the end of the Friday parent segment, but due to the cost and additional work on the final program day, in addition to the fact that some parents have a long drive home, we discontinued the snack/socializing segment of the closing session. Another former aspect of the program that incorporated parents was a year when we attempted to conduct a one-hour parent workshop on the opening and closing days of the camp. For the first session, we briefly noted the main topics the girls would work with and the instructional approaches we use and then had the family members engage in some tasks within those areas (e.g., nonroutine word problems, one of which required calculator use) and pentomino problems that exercised spatial skills and work with area and perimeter. The second session, held in the computer lab, provided an introduction to *Terrapin Logo* and *Geometer's Sketchpad.* Thirteen parents attended the first session and five parents the second. They trickled in throughout the hour. Although the instructor said some parents seemed to enjoy themselves, some seemed challenged and uncomfortable at times, especially those who did not speak English. Although we could have worked to improve these sessions, limited resources for conducting them caused us to abandon the idea.

Program Cost and Funding

The per-girl cost for this overnight camp that starts early Sunday evening and ends late Friday afternoon is approximately $1,000. This includes tuition, housing, meals and snacks, group recreation, camp t-shirt, math souvenir (a spatial puzzle), and other program costs. However, we seek and acquire funding annually—mostly from private foundations and local businesses—that help offset participants' registration fee. As of the 2016 camp, we charge $450 per camper and offer girls on free or reduced-price federal lunch status registration fees of $45 and $90, respectively. These reduced fees (compared with actual costs) allow more girls to participate in the program. Even then, the prices are a "stretch" for some families. We also like to cover all costs for the week to minimize socioeconomic differences among participants as much as possible. (SES markers still surface to some degree, however, by way of possessions and snacks that the girls bring to the camp.) In an effort to further financially support the girls who want to attend this camp, we provide accepted participants with a letter they may use to approach their school, businesses, individuals, and so forth, as they see fit, to request financial support for them to participate in the program.

Publicity efforts can encourage financial backing. We fax a press release a few days before each summer camp (longer than that has proved less effective) to university media as well as local newspapers and television programs and have notified the media about our new after-school computer programming club. The media are most interested in our OST efforts when we offer something new, such as a visibly engaging engineering or technology component or a special lunch session with a local female STEM professional seated at each table, or when we reach a milestone, such as our tenth anniversary. So, it is important to highlight these special program features in PR efforts, for which we enlist campus media relations personnel in addition to employing our own strategies. Conducting presentations to local organizations and at professional meetings, publishing articles and other literature that targets researchers and practitioners, and creating and maintaining a well designed and useful program website can further illuminate the program in the public eye.

Challenges

One issue faced with running a program such as this is getting the word out to all eligible girls and encouraging girls who might need the program most to participate. Despite the fact that we receive more applications than we can accept, we would like all girls to have an opportunity to consider taking part in the program. Although we distribute camp applications to all relevant schools across the northern part of the state, we have learned that many girls never receive them. We have tried variably sending them to school principals, counselors, and office staff, but ultimately getting the forms in the girls' hands is beyond our control. We also have been told that some teachers only distribute the applications to higher-ability girls, despite cover-sheet directions to give the form to *all* girls. Even when girls do get the application form, they might not give it to their parents or their parents might not be able to read it. (At some point we would like to translate the form to Spanish, the state's dominant second language.) When parents do receive and can read the application form (or their daughters can), some families might not recognize the importance of the program for their daughters or have the means to get girls to the camp in terms of cost, transportation, or need for potential campers to perform home duties, such as caring for younger siblings (Davis & Hardin, 2013). Indeed, youth from lower-income families have been shown to be less likely to participate in OST programs for these reasons and can thus fall farther behind more economically advantaged peers academically during the summer months (e.g., McCombs et al., 2012; Slates et al., 2012). Further, sometimes a few girls who sign up for the program do not show up and some withdraw early, usually due to homesickness or inability to adjust to the unfamiliar setting, which causes additional time investment (e.g., to replace the girls) during the busy camp week.

We experience a variety of other issues in running this program. A significant challenge is securing ongoing funding to offset participant costs. Although we have been successful with this to date, funding efforts require a great deal of work and cause some stress because this annual task involves the possibility of "tapping out" local support, making camp continuation uncertain on a year-by-year basis. Another issue that can be problematic is securing two university classrooms for instruction because summer courses have first priority for facilities. This also limits the possibility of expanding to three camper groups during the same week. (The computer lab and residence hall spaces are also needed, but we have been able to schedule these fairly successfully to date if we do so far enough in advance.) Some other issues include teaching a broad range of girls within each classroom (much more variable than a typical classroom that draws from a much smaller locale), handling behavioral issues with campers (in particular, relational aggression, accusations of theft, and use of cell phones, which are banned during camp), and getting the girls to make use of the Edmodo site embedded in the program website.

PROGRAM EVALUATION METHODS

Numerous evaluation measures have been used with program participants and their parents during this program's 18-year history, in particular, various types of surveys containing open-ended questions and statements to be rated on response scales, as well as small-group interviews. In this chapter, the data discussed come from survey ratings for one camp year and open-ended survey responses for two camp years.

Participants

Participants for the quantitative data reported here were 34 Nevada females ages 12–13 who attended the Northern Nevada Girls Math and Technology Camp in the summer of 2014. Qualitative data obtained as responses to open-ended questions came from 99 girls (64 rising seventh graders, 35 rising eighth graders) who attended the 2014 and 2015 camps and completed the camp-end evaluation.

Instruments

Mathematics as a Gendered Domain Instrument
Participants completed the Mathematics as a Gendered Domain (MGD) instrument as a measure of gender-related beliefs (Leder & Forgasz, 2002). The three subscales are Mathematics as a Male Domain (MD), Mathematics as a Female Domain (FD), and Mathematics as a Neutral Domain (ND). When administered, the 48 items of the three subscales were presented in random order. Participants indicated the extent to which they agreed or disagreed with each statement using a five-point Likert scale. Sample items for the three subscales include:

MD: Mathematics is easier for men than it is for women.
FD: Girls are more suited than boys to a career in a mathematically-related area.
ND: Girls and boys who do well in a mathematics test are equally likely to be congratulated.

The highest total score possible on this instrument is 240, with five points assigned to each item. The highest scores possible for the subscales are 80 for the 16 MD items, 85 for the 17 FD items, and 75 for the 15 ND items. Participants completed the instrument the first night of the camp, prior to the beginning of mathematics instruction, and again at the end of the camp.

Forgasz, Leder, and Kloosterman (2004) determined the construct and content validity for this instrument by reviewing research literature and seeking input from mathematics educators and students. Additionally, they report:

> A reliability analysis was conducted on the items comprising each of the three subscales of the *mathematics as a gendered domain* instrument. For each subscale, item-total correlations confirmed the internal consistency of the items. Cronbach's a [alpha] values for the three subscales were .90 for MD, .90 for FD, and .84 for ND. (p. 399)

Camp-End Evaluation Form

Near the end of the camp on the last day, the girls complete an evaluation form with items they rate on a five-point scale and others requiring written responses. The data shared here are derived from the following four open-ended questions on that form:

- Name up to three features of the program as a whole that you found most important.
- Name up to three features of the program's instructional approach that you found most important.
- If your schedule permitted and the program continued for your age group, would you be likely to attend the summer camp again next year?
- Offer any suggestions that might help us to improve this program, not only during the summer camp but also during the rest of the year.

Data Analysis

For the quantitative data, we used a paired-samples t-test to compare means for pre- and post measures, testing for significance at the .05 level using a one-tailed test. Students' beliefs toward mathematics, as measured by the MGD instrument, were the dependent variables. Of the 34 participants,

31 pairs of scores were analyzed for those who completed the instrument at both survey administrations.

We analyzed the qualitative data for themes. Through multiple readings of the written comments, we identified themes and adjusted them, as needed, until we believed that the themes reflected participant comments.

Results

Attitudes and Beliefs

Table 5.2 shows mean Mathematics as a Gendered Domain subscale scores for the beginning and the end of the camp. It shows that no significant difference appeared between the pre and post measures on the Male Domain or Neutral Domain subscales. However, significant favorable change occurred on the Female Domain subscale ($t = 2.78$, $p < .05$), indicating that participants perceived mathematics as a female domain to a greater degree on the second administration of the instrument.

Table 5.3 lists three items from the Female Domain subscale that achieved statistical significance between the pre and post measures.

Favorite Camp Components

When asked on the camp-end evaluation form what they considered to be the three most important features of the program, the girls focused mainly on the academic aspects of the camp. Almost 60% of the comments named specific subject areas or noted new learning and improved mathematics and technology skills. For example, one rising seventh grader wrote, "Learning math for future jobs. And improving the world," and a rising eighth grader listed "the problem solving because it made me think in different ways." Some additional comments (6% of all comments) also centered on the educational components of the camp in the form of valuing the session on learning biographical information about women in STEM and the student-centered instructional approaches. Next in line to

TABLE 5.2	**Mathematics as a Gendered Domain Subscale Scores**						
	Pre		Post				
Subscale	*M*	*SD*	*M*	*SD*	*t*	*p*	*d*
MD	31.90	10.57	33.39	12.37	.85	.201	.152
FD	43.55	11.17	49.23	14.23	2.78**	.004	.499
ND	61.90	8.15	62.39	9.15	.35	.364	.626

Note: A paired-samples *t*-test was used to compare means for the pre and post measures for 31 pairs of scores, testing for significance at the .05 level using a one-tailed test.
** $p < 0.01$

TABLE 5.3 Statistically Significant Differences in Items on the FD Subscale

Items	Pre		Post		t	p	d
	M	SD	M	SD			
28. Compared to girls, boys give up more easily when they have difficulty with a mathematics problem.	2.06	.96	2.52	1.15	2.04*	.025	.366
32. In a mathematics class with both boys and girls, girls tend to speak up more than boys.	2.52	.99	3.22	1.20	2.83**	.004	.508
44. Girls are more likely than boys to say mathematics is their favorite subject.	2.51	1.02	3.03	1.30	1.83*	.038	.328

Note: A paired-samples *t*-test was used to compare means for the pre and post measures for 31 pairs of scores, testing for significance at the .05 level using a one-tailed test. Of the 48 items on the MGD instrument, these three items were significantly more favorable from the first to the second administration of the instrument.

*$p < 0.05$, **$p < 0.01$

the educational aspects of the program were the social and recreational aspects, which garnered 13% of the comments. Examples include "meeting new people and becoming friends," "connecting with girls like me," and "fun activities after a long day's work." Another 6% of the comments commended the staff for being positive mentors and role models who they found encouraging, supportive, helpful, and accessible. Finally, almost 6% of the comments named spending time on a college campus—in particular, staying over night in a dorm—as a top camp feature.

Top Instructional Approaches

In listing up to three most important instructional approaches, the girls identified two main instructional features: active learning and the staff. Most of the 32% of comments that focused on active learning specifically named "hands-on." Unlike most other responses, a large difference appeared between rising seventh and rising eighth graders, with 38% of comments made by the former and 13% by the latter indicating the importance of active learning tasks. The 28% of comments that focused on staff foremost noted an appreciation for individual attention from staff (e.g., "one on one instructional help" and the staff's "direct work with *all* students") and for instructor effectiveness (knowledgeable, helpful, caring instructors), with having different teachers throughout the day being mentioned to a smaller degree. Almost 9% of comments listed group work, and about 6% described the expectation that the girls reason through and explain their work. In the latter case, for example, sample comments include the instructors "having people share thoughts" and "helping us figure something out how to solve it rather than just telling us." Instructional approaches named in 2–3% of comments each involved provision/creation of useful take-home materials (notebooks, handouts), effective teacher explanations (explaining things slowly and more than once), "fun" approaches, helpful tips and strategies, use of technology (e.g., graphing calculators and tablet computers) as a learning tool, and the supportive learning environment.

Other Evaluative Feedback

When asked whether they would attend the camp again next year if held for their grade level, participants overwhelmingly said yes. All but one of the 88 girls who responded to this question said yes. The only "no" response was "because I don't like spending the night." In response to an item soliciting suggestions for improving the program, many (20 girls) said none were needed (e.g., "Everything is perfect"). The main suggestion, offered by 13 girls, was to extend the program in length (e.g., two weeks) or to additional grade levels. Most of these comments (12) were made by the rising eighth graders, who expressed disappointment that they could not return to the program the next year. Only three other viable suggestions were made by a small number of girls: more time for sleep (6 girls, all rising seventh graders), more spatial-reasoning time (4 girls, all rising eighth graders), and more computer time (4 girls, 2 each for each age level). (*Note:* Comments were not tallied for camp policy-related requests, such as wanting to be able to have cell phones at the camp or to use elevators in campus buildings.)

DISCUSSION AND PRACTICAL IMPLICATIONS

The data reported here and in other publications about this program indicate that an OST mathematics and technology program for girls can favorably influence program participants in terms of dispositions and perspectives, as well as STEM performance and participation. In relation to findings detailed in this chapter, participants showed a significant favorable change in their perception of mathematics as a female domain but not in relation to mathematics as a male or neutral domain. Perhaps the indicated area of change is most important in that it might be better to see mathematics explicitly as female-appropriate than to see it as "not male" or even as neutral, which is more open to interpretation.

It is encouraging that the girls who participated in this program largely named academic program aspects as most important. This reminds us that girls (and students in general) can be "hooked" by and engaged in STEM academic content, even when challenging. We believe this is especially true when STEM material is presented in a comprehensible manner using active learning in a safe setting and when youth are helped to see the value of STEM to their lives and to the world at large. To a much lesser degree, but listed second in importance, were the social and recreational aspects of the program. Perhaps the marriage of the social/recreational with the academic enhances learning and motivation for girls because it exercises a fuller range of human characteristics that synergistically serve to strengthen each other.

The three items on which the girls showed significantly more favorable scores from camp beginning to end reflect program values that are mirrored in instructional approaches and program "messages." We explicitly discuss the importance of perseverance during challenge and the fact that struggle does not imply lack of competence. This experience is, we say, indeed normal during the course of worthwhile learning and that productive effort can lead to better performance (e.g., Barnett, Sonnert, & Sadler, 2014). We not only expect and encourage the girls in our program to persevere in their mathematical tasks, but we have them collaborate with peers and share ideas in small-group and whole-class settings. The girls are responsible to explain and defend their work to camp peers, and classmates are expected to comment on and pose questions about work shared publicly in the classroom. We emphasize a "Yay math!" and "We love math!" atmosphere throughout the week via explicit comments and enthusiasm toward the subject matter. Therefore, the girls' particular improvement in self-perceptions of females' perseverance, oral participation in mathematics class, and affinity for mathematics as a subject area demonstrates success in our efforts as well as the potential influence of programs designed with this type of pro-STEM orientation for girls.

In the data reported here and elsewhere for this program (e.g., Wiest, 2004, 2010), a dominant finding is that this summer OST STEM program for girls has improved participants' dispositions, especially their confidence. However, program evaluations in the form of surveys and interviews with the girls and their parents show that the program also appears to increase participants' knowledge, skills, and mathematics performance, as well as their participation in mathematics (e.g., elective opportunities such as clubs). Less prominent but other noteworthy results include greater participant awareness of college and career options, as well as personal and social benefits (e.g., improved skills in independent living and in making friends). The collective findings point to the program's academic content and its social and recreational aspects as being the most important program components. Other key program features reported in participant and parent program evaluations are the camp's high-quality staff, its instructional approach (mainly, use of group and hands-on work in a supportive, reasoning-oriented environment), and experience with campus life (staying in dorms and eating in the campus cafeteria). In the research reported in this chapter, we noted that a much larger proportion of rising seventh than eighth graders named hands-on learning as an important instructional feature. We believe this is because instructors for the younger group used this approach more often, thus bringing it to mind more readily. However, we contend that hands-on approaches (as long as they are used discriminately and as thinking aids) are important to and appreciated by girls of all ages.

CLOSING COMMENTS

Although the data reported in this chapter did not focus on confidence, based on other investigations of and almost two decades of experience with this program, we believe the confidence boost girls gain from participation in this program is the most important outcome. While increased knowledge and skills are also quite important, they are necessarily limited to the small subset of mathematics content that can be addressed in a finite amount of time, whereas confidence can linger and grow well beyond the program in an unbounded sense. This seems to be the root of what inspired many girls years after leaving the program, as discerned by many anecdotal comments shared by alumni or their parents. Improved attitudes can influence girls to continue learning mathematics both formally and informally, to participate in voluntary opportunities, and to advocate for themselves. While the high-quality staff and their effective, student-centered instructional approaches were reported as being very important aspects of the camp week, these are components that are beyond the control of the girls in their continued formal schooling but that have implications for the continued efforts of teacher educators and professional development staff.

Some aspects of the camp are conducive to good practice in mathematics teaching and learning in ways that traditional schooling cannot mirror but which support the value of OST programs because they can exercise greater freedom. For example, the number of staff in each of the camp classrooms affords greater individual attention than is possible in compulsory K–12 schooling. Camp staff have also noted that they may focus on more meaningful mathematics teaching in a no-homework, no-testing climate. Further, more intense, direct exposure to college life due to the location of the program can plant important ideas about pursuing a college degree.

Finally, the social and recreational aspects of the camp that the girls highly appreciated "rounded out" the experience in ways that support learning by providing needed mental breaks and connecting females interested in mathematics. They supplement the academic focus of the camp and could be experienced outside of the program. Nevertheless, we believe the social and recreational aspects helped foster the camper characteristic we contend matters most—confidence in one's ability to learn mathematics and to gain greater tools for doing so. The peer bonding and interaction with adult role models that occurred during social and recreational activities in the atmosphere of a shared mathematics-oriented experience also built confidence in the girls individually within the strength of the greater whole. Such a setting centered on females and mathematics can be more difficult to locate or create outside of this type of program.

Although only a few girls (four each) suggested incorporating more time for spatial-reasoning tasks and for computer work, these recommended improvements merit consideration. Both spatial and computer skills tend to be weak areas for females and are important to the STEM disciplines (e.g., Wiest, 2010), so more time and rigor spent with these skills might be vital factors in strengthening STEM competency and dispositions.

We believe that out-of-school-time STEM programs for girls are one effective way to address females' underrepresentation in STEM careers and to improve their abilities and dispositions related to using STEM in everyday life. Future research and evaluation should be designed to investigate outcomes of OST STEM programs for females, disaggregated by different identity groups (e.g., those based on family income level), and the critical program elements that generate those outcomes. Further, it is especially important to determine the sustainability of these self-funded programs and to use both quantitative and qualitative methods to assess the long-term effects of these programs on females' STEM performance and participation, as well as on their dispositions and the influence these dispositions might have on young women's career paths and personal lives.

AUTHOR ACKNOWLEDGMENT

The authors would like to thank the major program donors for the years in which the data reported in this chapter were collected: Odyssey Foundation; Newmont Mining Corporation; Frances C. and William P. Smallwood Foundation; International Game Technology (IGT); Barrick Gold of North America; AAUW Reno Branch. They would also like to thank the College of Education at the University of Nevada, Reno for providing facilities needed to conduct the program.

REFERENCES

Ashcraft, C., Eger, E., & Friend, M. (2012). *Girls in IT: The facts.* Boulder, CO: National Center for Women & Information Technology, University of Colorado. Retrieved from https://www.ncwit.org/resources/girls-it-facts

Augustine, C. H., & McCombs, J. S. (2015). Summer learning programs yield key lessons for districts and policymakers. *The State Education Standard, 15*(1), 11–19.

Barnett, M., Sonnert, G., & Sadler, P. (2014). Productive and ineffective efforts: How student effort in high school mathematics relates to college calculus success. *International Journal of Mathematical Education in Science and Technology, 45*(7), 996–1020.

Constan, Z., & Spicer, J. J. (2015). Maximizing future potential in physics and STEM: Evaluating a summer program through a partnership between science outreach and education research. *Journal of Higher Education Outreach and Engagement, 19*(2), 117–136.

Davis, K. B., & Hardin, S. E. (2013). Making STEM fun: How to organize a STEM camp. *TEACHING Exceptional Children, 45*(4), 60–67.

Dillivan, K. D., & Dillivan, M. N. (2014). Student interest in STEM disciplines: Results from a summer day camp. *Journal of Extension, 52*(1), 1–11.

Forgasz, H. J., Leder, G. C., & Kloosterman, P. (2004). New perspectives on the gender stereotyping of mathematics. *Mathematical Thinking and Learning, 6*(4), 389–420.

Goetz, T., Bieg, M., Lüdtke, O., Pekrun, R., & Hall, N. C. (2013). Do girls really experience more anxiety in mathematics? *Psychological Science, 24*(10), 2079–2087. doi:10.1177/0956797613486989

Grossman, J. M., & Porche, M. V. (2014). Perceived gender and racial/ethnic barriers to STEM success. *Urban Education, 49*(6), 698–727.

Heaverlo, C. A. (2011, January 1). STEM development: A study of 6th–12th grade girls' interest and confidence in mathematics and science. *ProQuest LLC.*

Johnson, C. C., & Walton, J. B. (2015), Examining the leaky STEM talent pipeline— Need for further research. *School Science and Mathematics, 115*(8), 379–380. doi: 10.1111/ssm.12148

Leder, G. C., & Forgasz, H. J. (2002). *Two new instruments to probe attitudes about gender and mathematics.* Retrieved from http://files.eric.ed.gov/fulltext/ ED463312.pdf

Lubienski, S. T., Robinson, J. P., Crane, C. C., & Ganley, C. M. (2013). Girls' and boys' mathematics achievement, affect, and experiences: Findings from ECLS-K. *Journal For Research in Mathematics Education, 44*(4), 634–645.

Lum, S. P. (2015). *Structure and agency in organizational contexts of women in STEM.* (Unpublished master's thesis). Michigan Technological University, Houghton, MI.

McCombs, J. S., Augustine, C., Schwartz, H., Bodilly, S., McInnis, B., Lichter, D., & Cross, A. B. (2012). Making summer count: How summer programs can boost children's learning. *The Education Digest, 77*(6), 47–52.

McGee, E. O. (2013). High-achieving Black students, biculturalism, and out-of-school STEM learning experiences: Exploring some unintended consequences. *Journal of Urban Mathematics Education, 6*(2), 20–41.

Mohr-Schroeder, M. J., Jackson, C., Miller, M., Walcott, B., Little, D. L., Speler, L., . . . Schroeder, D. C. (2014). Developing middle school students' interests in STEM via summer learning experiences: See Blue STEM Camp. *School Science and Mathematics, 114*(6), 291–301.

Mosatche, H. S., Matloff-Nieves, S., Kekelis, L., & Lawner, E. K. (2013, Spring). Effective STEM programs for adolescent girls: Three approaches and many lessons learned. *Afterschool Matters,* 17–25. Retrieved from files.eric.ed.gov/ fulltext/EJ1003839.pdf

Newbill, P. L., Drape, T. A., Schnittka, C., Baum, L., & Evans, M. A. (2015, Fall). Learning across space instead of over time: Redesigning a school-based STEM curriculum for OST. *Afterschool Matters*, 4–12.

Paulsen, C. A. (2013, Spring). Implementing out-of-school time STEM resources: Best practices from public television. *Afterschool Matters*, 27–35.

Riegle-Crumb, C., King, B., Grodsky, E., & Muller, C. (2012). The more things change, the more they stay the same? Prior achievement fails to explain gender inequality in entry into STEM college majors over time. *American Educational Research Journal, 49*(6), 1048–1073.

Ross, J. A., Scott, G., & Bruce, C. D. (2012). The gender confidence gap in fractions knowledge: Gender differences in student belief-achievement relationships. *School Science and Mathematics, 112*(5), 278–288.

Shapiro, M., Grossman, D., Carter, S., Martin, K., Deyton, P., & Hammer, D. (2015). Middle school girls and the "leaky pipeline" to leadership. *Middle School Journal, 46*(5), 3–13.

Slates, S. L., Alexander, K. L., Entwisle, D. R., & Olson, L. S. (2012). Counteracting summer slide: Social capital resources within socioeconomically disadvantaged families. *Journal of Education for Students Placed at Risk, 17*(3), 165–185.

White, D. W. (2013). Urban STEM Education: A unique summer experience. *Technology and Engineering Teacher, 72*(5), 8–13.

Wiest, L. R. (2004). Impact of a summer mathematics and technology program for middle school girls. *Journal of Women and Minorities in Science and Engineering, 10*(4), 317–339.

Wiest, L. R. (2010). Out-of-school-time (OST) programs as mathematics support for females. In H. J. Forgasz, J. R. Becker, K.-H. Lee, & O. B. Steinthorsdottir (Eds.), *International perspectives on gender and mathematics education* (pp. 55–86). Charlotte, NC: Information Age.

CHAPTER 6

GREATER OPPORTUNITIES ADVANCING LEADERSHIP AND SCIENCE (GOALS) FOR GIRLS AT THE INTREPID SEA, AIR & SPACE MUSEUM

**Jeanne Houck, Lynda Kennedy, Sheri Levinsky-Raskin,
Kerry F. McLaughlin, and Shihadah Saleem**

ABSTRACT

The Intrepid Sea, Air & Space Museum's GOALS (Greater Opportunities Advancing Leadership and Science) for Girls is a year-long, multifaceted program intended to engage New York City teens with STEM concepts and careers. The cornerstone of the program is a six-week, full-day intensive summer program at the museum for 50 rising ninth- and tenth-grade girls, followed by year-round weekend STEM programs and an internship opportunity. Program outcomes are measured using a range of qualitative and quantitative information-gathering methods, such as pre- and post-program surveys and observations investigating knowledge, attitudes, and behaviors relevant to each project component. The findings show: (a) increased knowledge of science topics engaged; (b) improved science and leadership skills; (c) broader

Out-of-School-Time STEM Programs for Females, pages 123–149

awareness of career opportunities in the sciences; and (d) greater confidence, personal awareness, and understanding of the value of teamwork and the role of leadership in creating innovations.

The Intrepid Sea, Air & Space Museum's Greater Opportunities Advancing Leadership and Science (GOALS) for Girls program, which began program delivery in 2008, is a year-long program that aims to engage New York City high school girls with STEM subjects and careers in order to build and support a pipeline of talent. The program provides resources for learning STEM subjects and exploring college and career paths, along with opportunities to meet women in STEM careers, arrange internships, and build a social/professional network. The cornerstone of the program is a six-week, full-day Summer STEM Intensive for 50 girls before their ninth- or tenth-grade school year. Each Summer Intensive cohort is then invited to attend GOALS Forums on weekends during the school year with their families. Forums also include GOALS alumnae from previous years and participants from other organizations working with girls in STEM.

That GOALS has become a core program of the museum's education offerings is not surprising. Founded in 1982, the Intrepid Sea, Air & Space Museum is a non-profit institution focused on education. The overall mission of the museum is to "promote awareness and understanding of science, history, and service in order to honor our heroes, educate the public, and inspire our youth." Centered on the historic aircraft carrier *Intrepid*, a national historic landmark, the museum's unique collection spans a half-century of technological innovation. Highlights of the collection are 27 aircraft, including the A-12 and the British Airways Concorde, the Cold War-era submarine *Growler*, and the space shuttle *Enterprise*. As home to NASA's largest artifact in the Northeast, the museum is a focal point for interpreting the importance of NASA to past and future American aviation and international spaceflight.

The Intrepid Museum is also an unconventional classroom. Each year participants in the museum's myriad education programs learn about science, technology, engineering, and mathematics (STEM) subjects through technological and scientific innovations in aviation, seagoing vessels, and space exploration, which are represented by the museum's major artifacts. Education programs at the museum situate advances in scientific and technological innovations within historical contexts that drove the need for creative problem solving. The museum education staff conducts STEM programs with an emphasis on innovations in engineering with more than 30,000 participants each year, more than 12,000 of whom participate at no charge.

An important goal for the Intrepid Museum staff is making museum resources and experiences free or affordable to all. In New York City, more than one million students attend approximately 1,600 public schools, many of which are considered high-need schools in low-income neighborhoods. Studies show that only 43% of low-income students visit a museum by the end of their kindergarten year, compared to 65% of high-income students, and this disparity persists into the upper grades (Institute of Museum and Library Services, 2013). According to ExpandED Schools (n.d.), children from middle-class backgrounds are likely to engage in 6,000 more hours of learning opportunities (e.g., participation in after-school and summer programs) by sixth grade than those from low-income backgrounds. Gaps in educational experiences are particularly high during the summer, leading to "summer slide," and many low-income students lack access to an affordable, quality option for summer programs. Expanding access to Intrepid Museum resources for all students and families, regardless of ability to pay, is central to the goals of the museum's education team. The program that is the focus of this chapter, the Greater Opportunities Advancing Leadership and Science (GOALS) for Girls, grows directly out of a desire to promote equity and access for all.

CONTEXT

As schools continue to face budget and other constraints limiting exposure to STEM content, the Intrepid Museum's GOALS for Girls program helps to fill program and curriculum gaps in girls' STEM education. According to a 2011 U.S. Department of Commerce report, women are underrepresented nationally in STEM undergraduate degrees and jobs. Although it is difficult to pinpoint the exact causes of this underrepresentation, the report concludes that key factors might be stereotypes that situate men in STEM careers and that accrue from a lack of female role models in STEM fields. The GOALS for Girls program addresses these challenges by immersing young women in engaging STEM curricula and activities, fostering self-confidence in their ability to develop and share knowledge, and helping them envision science careers by exposing them to female professionals in STEM fields. While the GOALS program structure varies from component to component, the target results for the full repertoire of GOALS programming are:

- To broaden student perspectives on STEM topics and careers
- To provide opportunities for students to practice problem-solving and critical-thinking skills
- To prepare students for subsequent STEM programming and educational pursuits, and ultimately STEM careers

Objectives for GOALS for Girls closely align with those recommended in recent studies for addressing the academic and professional gender gap in which women are underrepresented in science. For example, the program is designed to influence participants long-term (McCreedy & Dierking, 2013). McCreedy and Dierking (2013) studied six STEM programs for girls by communicating with women who had participated in the programs 5–25+ years earlier. The findings indicate that "memories of STEM experiences became critical resources in girls' stories about their lives" (p. 32) and that informal STEM experiences were opportunities for participants to explore and to continue engaging in STEM-related experiences.

Although the core of the GOALS program is rigorous STEM content, the program also aims to help participants develop relevant analytical and socio-emotional skills. These goals align with recommendations by the White House Office of Science and Technology Policy (2014) regarding needed 21st Century skills in STEM education. The GOALS program also promotes key skill areas identified by the Partnership for 21st Century Skills (n.d.), including learning and problem-solving skills; information, media, and technology skills; life and career skills; and 21st Century themes, such as environmental literacy.

PROGRAM DESCRIPTION

The GOALS for Girls program is a prime example of the Intrepid Sea, Air & Space Museum's dedication to promoting equity and access for all youth and of its efforts to contribute talent to the STEM pipeline from underrepresented groups. Funded through institutional commitment and the generous support of foundations and individual donors, all aspects of the GOALS experience are provided at no cost to participants.

First offered in 2008 in partnership with a network of girls schools and opened to girls from anywhere in New York City in 2009, enrollment in the GOALS for Girls Summer Intensive has become a competitive, application-based process that begins in January. Each year, the Summer Intensive is advertised to the New York City school community at large and through several other avenues, including the National Girls Collaborative Project and the Intrepid Museum's website, which is updated every year with information, downloadable forms, and statistics from previous years. Girls in eighth or ninth grade during the academic year are eligible for the Summer Intensive. Interested participants submit a rigorous application that includes closed-format items, short-answer questions, and a 500-word essay, as well as school transcripts and a recommendation from a mathematics or science teacher. For the application questions, the girls state who they admire most and why, list potential influencers from given choices to note who inspires the girls most in science (includes "other" for write-ins), write short answers about what they like most and least about STEM subjects, complete sentences such as, "When I do math, I feel _____," explain what comes to mind when they think of a scientist, engineer, and other STEM professionals, rate their confidence level in STEM subjects, and provide some basic demographic information. The questions are designed to give museum staff some baseline data and a sense of the personalities of the applicants for the staff members to consider in choosing a diverse and well-functioning cohort of girls. The application essay is based on one of three prompts. The options for the 2015 application were:

Essay Topic #1: Name one controversial issue in the scientific or technological world. Explain why it is disputed and how you, as a future scientist or engineer, would solve this dilemma. Provide at least two examples of ways you would solve it.

Essay Topic #2: If you could design your own exhibit at the Intrepid Sea, Air & Space Museum, what topic in STEM and history would you focus on? What types of interactives would you include to engage and educate visitors and why? Explain and describe the purpose of your exhibit and how it fits into the unified message of the museum.

Essay Topic #3: Explain how the quote below relates to your life. Give two examples of how you can or will clear hurdles for yourself and other girls interested in STEM, or how you already have.

> If we're going to out-innovate and out-educate the rest of the world, we've got to open doors for everyone. We need all hands on deck, and that means clearing hurdles for women and girls as they navigate careers in science, technology, engineering and math.
>
> —First Lady Michelle Obama, September 26, 2011

A teacher recommendation is submitted as a form that asks the teacher to rate an applicant in several areas, such as work habits and social skills, and provides an opportunity to write a more personalized recommendation. Transcripts or report cards are required only for the current school year.

After narrowing the applicant pool, approximately 80 girls are interviewed. Museum staff carefully acknowledge and appreciate different personality types and are aware that the GOALS interview is usually the first formal interview these girls have experienced. The girls are asked to discuss past out-of-school-time experiences as well as answers on their applications in greater depth. Staff are interested to see if this would be the first full summer experience the applicant would have with just other girls. Many of the girls do not have an opportunity to engage in a girls-only experience focusing on academics, social, and professional skills on a level playing field outside of the school setting where they might think they are already labeled an "A" or a "C" student. Being shy or an English language learner or not being a particularly strong student in school does not knock girls out of the running for a place in GOALS. It simply becomes information to consider when thinking about the balance of the group and the best support for the student. The most important traits sought are a willingness to work hard and curiosity. All things being equal between two girls, the student who has had less opportunity or who may be from a less resourced school district is favored for acceptance. Although target participation is 50 girls, 55 are invited to the program. This strategy has been successful. Each year out of the approximately 55 girls who are accepted, 48–50 attend and complete the program.

The key component of the GOALS for Girls program is the GOALS Summer Intensive, formerly called Camp GOALS for Girls. Its new name reflects the serious academic focus and rigor of this six-week summer program. The Summer Intensive provides an interdisciplinary approach to exploring STEM content, providing participants with broader understanding of STEM subjects. The program is staffed entirely by women on the museum staff, augmented by women brought on as staff for the summer program. It takes place on a school day schedule (generally 9–3, with longer days as needed for a special experience), Monday through Friday for the six weeks. The "home base" for the summer is the Michael Tyler Fisher Center for Education of the Intrepid Sea, Air & Space Museum. The GOALS for Girls program is housed in a large, sunny, flexible space that staff can divide and use according to the task at hand, although the girls are often out and about in other areas of the museum or making site visits to topic-related organizations. Because the museum wanted to break down possible barriers to participation, the girls are given metrocards to cover bus and subway fares, and healthy boxed lunches are served daily.

In-depth program evaluation that includes surveying all participants after each unit informs program revisions. In their feedback, the girls expressed a desire for more time with subject areas. Accordingly, the model for the six-week program has changed from week-long units to three extended units of two weeks each in the following areas: (a) aerospace and space science, (b) environmental and earth science, and (c) technology and engineering. Each unit involves participant-driven projects and presentations, long-term investigations, individual journal reflections, and discussions and debates centered on thematic units of study. The girls are exposed to many STEM resources throughout the city while on field trips to off-site venues, such as labs, environmental centers, and museums. Because the GOALS for Girls program focuses not only on STEM content but also on supporting a pipeline of talent into STEM studies and careers, participants also have opportunities to build professional networking skills. GOALS girls practice interpersonal communication through networking workshops, formal and informal discussions, and contextualized role-playing. On the Friday that ends the second week of each unit, a Youth Leadership Conference (YLC) is held. YLCs provide the girls with an opportunity to engage with invited influential women in STEM fields in small and large groups. Female mentors whose expertise and experience fit the bi-weekly themes present their work and serve as important role models that are both accessible and relatable. These women, who include undergraduate and graduate students, postdoctoral fellows, professors, professionals, and educators from other STEM-based organizations, provide valuable insights and diverse perspectives. These early interactions increase awareness and offer support for girls to envision themselves in and pursue a STEM career. The girls are encouraged to practice dressing as if for an interview on these days and practice networking with the visiting professionals.

The first unit of the summer begins with time devoted to GOALS participants learning about each other. This is essential, as the participants are diverse in many ways and come from all over New York City. As noted, a strong effort is made to ensure GOALS serves girls who would not have a similar opportunity elsewhere. Approximately 75% of the schools the girls attend are designated as Title I (schools with at least 40% of the student body eligible for federal free or reduced-price school lunch) and are located in high-poverty areas. For the 2015 GOALS Summer Intensive, 35% of the participants self-identified as African American or Black, 20% Hispanic or Latina, 18% Asian, 8% White, and 18% multiracial/other. Three-quarters of the girls or their parents had newcomer status: 14% were born in other countries and 61% were first-generation with parents from a total of 21 countries. More than 70% of the girls reported that they had not previously participated in any STEM-related extracurricular activity. Creating a safe space with norms co-created by the girls themselves and mutually accepted (see example in Figure 6.1) is the first step to encouraging participants to move out of their comfort zone and to take risks. It also begins to build a group identity as GOALS girls—that is, girls who do STEM—and who will continue to support each other as they move through high school and perhaps beyond.

The first round of STEM content involves the physics and science behind aviation and space exploration. The museum's collection of 27 aircraft is the focal point, and the reality of the Intrepid as an aircraft carrier illustrates in physical space the history and engineering behind aviation innovations. By incorporating the museum's exhibits that highlight the NASA space shuttle Enterprise, the Soyuz TMA-6 spacecraft, the Intrepid's ties to NASA's Mercury and Gemini missions, and material from the museum's annual Space and Science Festival, participants have an opportunity to

Figure 6.1 Co-created and mutually accepted norms established for the instructional setting.

explore a breadth of topics about aerospace engineering and astronomy. The culminating gathering on the last Friday of the unit includes engaging conversations with influential female pilots, engineers, scientists, astronomers, and astronauts.

In the second unit of the summer, which focuses on the Hudson River, the girls explore marine and earth science by studying the river's health, reviewing the human impact on its environment, determining the physical interactions of ocean currents, and debating global warming. Guest workshops conducted by environmental organizations provide varied perspectives on habitat destruction and preservation. The girls visit the museum's submarine to gain additional historical and technological perspectives about engineering for meeting the challenges of long-term living at sea and visit local institutions with related content. They learn marine and environmental science through an interdisciplinary and holistic approach that employs varied methods. For example, during a visit to Brooklyn Bridge Park Conservancy, the girls were immersed in the city's natural habitat by learning to kayak and sail. Many city-dwelling youth have never had the opportunity to learn to sail, kayak, or interact with the natural environment in safe and informative ways. Educators from the South Street Seaport Museum helped the girls learn about the history of New York City waterways and the ecology of New York Harbor and to conduct on-site water quality tests. These immersive field trips allow program participants to gain deeper understanding of the natural world and the impact of human population growth, industry, and urban expansion on these delicate environments.

During exploration of environmental science, human impact, biological perspectives, and cognitive discoveries, partners such as Rockefeller University's DNA Lab and NY Hall of Science have helped GOALS girls make connections between DNA and diseases. The girls participate in a brain dissection lab, giving many their first dissection experience. Educators with live collections have come in to discuss animal diversity and teach respect for the natural world through interactions with live animals. The culminating Youth Leadership Conference on the last Friday of this unit includes engaging conversations with influential female environmentalists, biologists, and marine scientists.

During the final unit of the Summer Intensive, the girls participate in engineering design challenges through various activities and workshops. They are immersed in real-life applications and processes that integrate key concepts of engineering and marketing. TinkerCad software and 3D printers are used to create participant-inspired, engineered prototypes of week-long projects. After a period of guided research, the girls present their work in a large-group setting. Examples of past projects have included a robotic hand with working digits capable of grasping and turning a door handle, building a structure that could bear a girl's weight, and creating a communication system for a small settlement. GOALS for Girls Summer Intensive participants have also been invited to the Cooper Union School of Engineering for their summer STEM high school student presentations that provide opportunities to engage with peers and hear from high school sophomores and juniors immersed in a six-week engineering program. The culminating Youth Leadership Conference on the last Friday of this unit includes conversations with female engineers and designers in environmental science, computer science, marine science, and auto mechanics.

The Summer Intensive program concludes with two special nights during the final week. The first is a family open house in which families are invited to the museum to witness the girls' successes in the Summer Intensive. It also allows the girls to display personal skills and talents in a talent show. GOALS participants plan the content of the open house, including which work will be displayed, the content and schedule of the talent show, which girls will serve as family greeters, and who will be the emcee for the talent show. The importance of inviting families to see the pride, confidence, and knowledge their daughters have gained is supported by research indicating the importance of extended networks, such as families, on the persistence of girls and women in STEM pursuits (Modi, Schoenberg, & Salmond, 2012; Munley & Rossiter, 2013). Museum staff has also begun to engage GOALS girls' families through the Summer Intensive Tumblr page, which regularly communicates the girls' accomplishments while they are participating in the program.

The second special night during the last week is an overnight experience where GOALS participants sleep at the museum. This is a fun but purposeful experience intended to enhance participant bonding that has taken place during the six weeks of the program. The morning before the sleepover, following mentorship and networking at the final Youth Leadership Conference with mentors, the girls have an opportunity to explore the museum in small groups, re-visiting areas of interest or going for the first time to areas not incorporated into the program. Following their self-guided exploration of the museum, participants and staff continue bonding and sharing experiences over a catered buffet-style dinner, while enjoying a slideshow of the past weeks. This is followed by a participant-voted, staff-approved movie with snacks, which tends to be the highlight of the evening. Additional evening activities, such as star gazing on the ship's deck, offer the girls more ways to expand their STEM thinking. The morning following the sleepover, participants are congratulated over a catered breakfast with an informal certificate-of-completion award ceremony. After tearful and happy goodbyes and "yearbook-like" signing of program journals and t-shirts, the GOALS for Girls Summer Intensive comes to a close. The overnight experience is an important way to culminate a robust six-week program, full of social and academic rigor, self-exploration and discovery, and development of tools and resources that support a promising career in STEM.

AFTER THE SUMMER

GOALS Forums continue STEM learning, mentorship, and college/career awareness into the school year for Summer Intensive alumnae, as well as their families and friends, in full-day weekend programs focused on select topics. Formerly known as Science Weekend programs, the name has been changed to GOALS Forums to reflect a wider variety of science-oriented and youth-development topics addressed. Annually, these programs serve approximately 200 girls as primary recipients drawn from GOALS alumnae and other STEM programs for females. The programs also serve an additional 200 individuals as secondary recipients, including families that join the girls in the programs and museum visitors who take part in program activities. The Girls in Science and Engineering Day in March draws over 1,400 participants, including members of the general public.

Some GOALS Science Forums provide hands-on, experiential lessons and activities that continue to focus on the three areas of study introduced in the Summer Intensive. Components might include a guest speaker to start the day and to inspire the girls and their families, followed by breakout sessions for the hands-on components. Girls in Science and Engineering Day is a museum-wide festival that invites like-minded organizations to present their programs to museum visitors. This event promotes relationships between the Intrepid Museum and other STEM organizations and raises awareness of STEM resources for young people that are available in the city. In addition to staff from these organizations, interns drawn from the GOALS program—called GOALS Navigators—present programs they developed during their internships (more information to follow).

Two GOALS Student Development Forums provide girls with the resources they need to determine the next steps in their academic or career paths. One forum, a college fair, includes interaction with 10–14 college representatives, as well as mini sessions designed to inform the girls about the application process for student aid and provide insight into student life at college. The second forum is Mentorship Day, another opportunity to provide inspiration and information about STEM careers. The format includes a keynote address to introduce the day, provide context, and inform girls ages 13–18 of the important steps needed to prepare for a STEM career. The day incorporates round-robin sessions between participants and mentors in intimate and friendly settings, allowing for inquiry and discussion about STEM careers and the academic paths mentors pursued in order to be successful in their careers. GOALS Forums also include opportunities to work in the Education Department's computer lab, which hones skills in upward-trending career pathways of coding, programming, graphing, and creating virtual simulations. Computer lab programs have included using basic coding resources, such as Scratch, and 3D computer-assisted design using Tinkercad, as well as exploring Thingiverse and 3D printing. Curriculum materials are adapted from existing resources, such as Creative Computing, Code.org, and the Google CS First site.

Mentioned earlier, the more recent GOALS for Girls Navigator Internship is an opportunity for GOALS Summer Intensive alumnae to continue their GOALS experience and build workplace skills. Navigators continue to learn STEM content while facilitating programming associated with GOALS and programming that serves the general public of the museum. Due to the popularity of the Navigator Internship, which in the 2014–15 school year had a maximum of 15 places, the program has expanded to accept 20–30 Summer Intensive alumnae as short-term interns. Each Navigator is required to work 80 hours and is paid a stipend. To be eligible for the Navigator Internship, girls must be in good standing with the program, submit an application, which includes short-answer questions and an essay, and participate in a group interview. GOALS alumnae ages 14–20 are eligible for the internship. Similar to the process for being accepted into the GOALS program, the answers and interviews in the application process for the Navigator internships are used to create a strong, well-functioning team made up of alumnae from several previous cohorts.

Through increased program hours and engagement with peers, interns have a more robust academic and professional portfolio and develop their interpersonal, teamwork, and networking skills. During the internship, Navigators collaborate in working groups and participate in a number of GOALS program elements. Once they are done with their internships, Navigators can apply to become part of the Intrepid Teen program, a paid, part-time, junior-educator position that draws from GOALS interns and interns from the Youth Leadership Institute @Intprepid program, which is a co-ed leadership program for high school juniors run by the museum.

KEEPING PROGRAM CONTENT CURRENT

With the ever-changing landscape of scientific and technological advancement, the STEM Advisory Committee is an integral part of establishing current, relevant content and skills for the GOALS program. The ten-member committee includes women who are engaged in advanced STEM study or STEM careers. As advisors, they serve as resources and provide museum staff with insight into changing trends in science and advice on developing lessons for STEM programming. They also serve as mentors to the Navigator interns.

PARTICIPANT DIVERSITY: CHALLENGE AND OPPORTUNITY

The GOALS program accepts diverse girls into its Summer Intensive. In 2015, 61% of the girls were first-generation American with parents born in a total of 21 countries, 76% had not participated in after-school STEM programs prior to GOALS, and of those who responded, 20% of their mothers had graduate and/or professional degrees, 35% completed college, and 37% completed some college. These data are examined to direct outreach practices, for example, to ensure that the program reaches girls attending school in high-poverty areas and to provide insight into the types of supports these young women need to pursue the sciences. The data further illustrate the rich demographic makeup of New York City girls who are eager to participate in STEM programming.

Participants are diverse culturally and ethnically, but also academically and socially. It is not unusual for girls to tend to gravitate toward groups to which they perceive themselves belonging and not extend themselves beyond their comfort zone. To combat this, GOALS staff members meet at the end of the day to discuss their observations of the girls' social and academic interactions. They also use this time to create groups/pairs for the following day's or week's activities.

The academic and/or social maturity of the girls that enter the program can vary quite a bit. Some girls might have greater STEM knowledge and skills, whereas others might have better developed social skills. The combination of both types of girls in this program allows higher-achieving girls to help elevate the performance of those with less advanced skills and understanding, while those who have stronger social skills can help "draw out" the more introverted girls. Mixing these types of girls illustrates that everyone has different strengths to contribute to a task.

PROGRAM EFFECTIVENESS

Evaluation Design

Coinciding with a newly established, dedicated staff member to oversee evaluation measures at the museum, the 2015–2016 GOALS year marked the beginning of a more targeted and comprehensive evaluation strategy for the program. Development of a logic model, an internal guide that visually outlines program outcomes, activities, and data collection processes, formalized the thinking behind each GOALS program element and clearly identified its intended outcomes for participating girls. (See Figure 6.2.) A variety of new data collection instruments and analysis approaches initiated creative methods for tracking and documenting changes in the girls' behaviors, skills, and interests.

Historical Background: Data-Based Decisions

Transformation of the GOALS program since its inception in 2008 and initial summer in 2009 are relevant to understanding decisions made over time about the program that involve, for example, its structure and rationale. Data influenced decisions in the program's early years, and while not as robust and comprehensive as that compiled since 2015, it included current research findings, trends in STEM education for girls, and snapshots of other local programs serving the same audience.

The first iteration of the GOALS program in the 2007–2008 school year began while the Intrepid Museum had closed for restoration. In partnership with five New York City public schools in the Young Women's Leadership Network, the GOALS for Girls program began as a series of science-based professional development programs led by Intrepid education staff for seventh- and eighth-grade teachers to learn to offer in-school programs

GOALS for Girls – Logic Model

INTREPID
SEA, AIR & SPACE MUSEUM COMPLEX

Outcome 1: To broaden student perspectives in STEM topics and careers.						
Indicator(s)	Applied to	Output	Information needed to determine if outcome is being attained	Data Source	Data Interval	Target
Students will build upon prior knowledge, increase and deepen academic experiences, and be exposed to the interconnectedness of sciences fields. Students will identify and share prior knowledge and articulate new areas of interest in STEM. Students will demonstrate increased awareness and relevance of the interconnectedness of the sciences. Students will articulate new understanding about science as shared responsibility.	**Summer Cohort; Alumnae attending weekend Science Forums; Navigator interns**	➤ daily activities in summer program with focus on varied STEM topics during each of the 3 units; some led over the course of sequential days • daily on-site workshops • at least 4 off-site field studies/workshops • guest presenters ➤ weekend science forums for alumnae ➤ Navigator internship schedule of sessions	1) student attendance and participation 2) what students already know and what they want to know 2a) student misconceptions about the STEM topics introduced 3) new / increased knowledge 3a) student interactions during programming, with each other, and with women from the field / workshop presenters	1) attendance records and schedule 2) Participant Perception Indicator (PPI) for level or experience and confidence regarding STEM topics 2a) crowd sourcing for student misconceptions, newly learned concepts, surprises 2b) journal entries; student application responses; student/staff meetings 3) completed activity worksheets; completed unit workbooks 3a) observation rubrics for activities and presentations	1) daily attendance 2) PPI on first day, mid-summer, and last day; beginning, middle, and end of Navigator internship 2a) crowd sourcing at the beginning and end of each unit in the summer; when new topics during summer units are introduced; at weekend summer forums 2b) weekly journal entries; applications; student/staff meeting schedule 3) worksheets collected at the end of activities; workbooks collected at the end of each summer unit during scheduled observations of group work each week 3a) select summer activities observed by staff; 3 round table end of summer unit observations; Navigator programming; science forum programming dates	A) At least 95% (47) of students in summer: • attend all activities, off site field studies, and guest presentations • self-report that their confidence and experience levels increased after the summer programming B) at least 95% (47) of summer cohort students and Navigator interns articulate (verbally or written) at least one newly learned concept and one new area of interest for them in STEM from the programming C) all Navigator interns present final projects with a focus on STEM topics and reflect their research

Figure 6.2 Logic model for GOALS for Girls' outcomes. *(continued)*

GOALS for Girls - Logic Model

INTREPID SEA AIR & SPACE MUSEUM COMPLEX

Outcome 2: To provide opportunities for students to practice problem-solving and critical thinking skills.

Indicator(s)	Applied to	Output	Information needed to determine if outcome is being attained	Data Source	Data Interval	Target
--- Students develop the necessary tools to become proficient in professional, social and academic setting and be better prepared for college and careers. Students will effectively communicate needs, develop team building skills, and be supportive of one another. Students will demonstrate positive and open body language, and improved communication and presentation skills.	Summer Cohort; Alumnae attending weekend Science Forums; Navigator Interns	➤ opportunities for individual and group work in summer program • individual presentations • group presentations • end of unit round table groups ➤ opportunities to meet women in STEM fields in the summer program • Leadership Conferences • workshop presenters • off-site field studies ➤ weekend science forums for alumnae • group work on activities ➤ Navigator internship schedule of sessions • public speaking • work experience in Museum offices and on programs	1) student attendance and participation 2) # of students demonstrating critical thinking skills 2a) # of students effectively communicating in their small groups to solve a challenges 3) student explanations and presentations 4) student appearances and behavior during presentations to each other, to the public, and with experts / presenters / women from the field 5) student interactions in groups and with each other and with women from the field / presenters	1) attendance records 2) staff observations via formal rubrics and information notes 2a) staff observations via formal rubrics and informal notes 3) journals; group work documentation 3a) group lists; crowd sourcing data; journal entries; student/staff check-in forms 4) staff observations via formal rubrics and informal notes 5) staff observations via formal rubrics and informal notes	1) daily attendance 2) scheduled group work; weekly small group presentations 2a) scheduled group work; specific topic focused group challenges; scheduled public presentation dates 3) weekly journal entries; scheduled written prompts 3a) scheduled group work; specific topic focused group challenges; scheduled public presentation dates 4) scheduled group work; specific topic focused group challenges; scheduled public presentation dates 5) scheduled group work; weekly small group presentations; meeting schedule	A) at least 85% (42) of the students comfortably shift between leader and non-leader roles during small group projects and presentations. B) At least 80% (40) of the students in the summer will develop better communication and public speaking skills. C) All Navigator interns demonstrate improved communication and presentation skills.

(continued)

Figure 6.2 (continued) Logic model for GOALS for Girls' outcomes.

GOALS for Girls – Logic Model

INTREPID
SEA, AIR & SPACE MUSEUM COMPLEX

Indicator(s)	Applied to	Output	Information needed to determine if outcome is being attained	Data Source	Data Interval	Target
Outcome 3: To emerge from the program prepared for their futures in subsequent STEM programming and educational pursuits.						
— Students will ultimately continue in STEM studies and onto STEM careers. Students will inquire about and/or participate in additional GOALS and community STEM-related opportunities. Students' perceptions about their involvement in STEM studies and careers will change. Students will encourage other girls to participate. Students will connect with new networks within the STEM community.	**Summer Cohort; Alumnae attending weekend Science Forums; Navigator interns**	∧ In all elements of the program • opportunities to ask thoughtful questions and speak with women from various STEM fields about their career paths and academic/extracurricular program choices • goal-setting group work • group work to define and compare professional and social behaviors ∧ weekend science forums for alumnae • meet women in STEM fields – guest presenters and workshop facilitators • opportunities for alumnae to invite friends and family to programming ∧ Navigator internship schedule of sessions • opportunities to attend professional networking programs and workshops • opportunities to be paired with a mentor and work with experts in STEM fields	1) student attendance at GOALS weekend science forums 2) student's personal, academic, and professional goals 3) level of knowledge about the number of different STEM careers 4) knowledge of colleges that offer STEM degrees within tristate area 5) level of portfolio upkeep 6) referrals by GOALS alumnae 7) alumnae participation in STEM networks and other programs / workshops	1) attendance records 2) self-reflection essay 3) Participant Perception Indicator (PPI) 4) Participant Perception Indicator (PPI) 4a) crowd sourcing 5) portfolio 6) registration lists and attendance 7) student updates – email, Google+ Page, requests for recommendation letters for applications to other programs	1) at each science forum program 2) navigator intern scheduled journal entries 3) PPI on first day, mid-summer, and last day 3a) beginning, middle, and end of Navigator internship 4) PPI on first day, mid-summer, and last day 4a) beginning, middle, and end of Navigator internship 5) after meeting women from varying STEM fields / presenters / workshops 6) at each science forum and subsequent summer applications 7) regular email correspondence with alumnae and Navigator interns 7a) staff and student updates to google+ Page.	A) At least 90% of the students from the previous summer attend 1/2 of the offered number of weekend science forums. B) Students are comfortable to speak candidly about their program experiences and future needs. C) Navigator interns are referring girls from their schools or other extracurricular programs to GOALS. D) Navigator interns report that GOALS provided them with necessary skills, experiences, and information that led to their involvement in subsequent STEM programming.

Figure 6.2 (continued) Logic model for GOALS for Girls' outcomes.

for their students. The museum reopened in the fall of 2008 and quickly adapted the program in response to research about science programs for girls and summer learning loss. Studies indicated that too few out-of-school-time programs for girls interested in science existed and, as Cech (2007) notes, closing the achievement gaps that plague our schools will not be possible until summer learning opportunities for disadvantaged youth becomes a policy priority and these opportunities are made accessible to all. The museum intentionally shifted the program's timing from the school year to the summer months. This led to the next iteration of the GOALS program in the summers of 2009 and 2010.

As the museum prepared for the summer of 2011, another deliberate change was made, this time to the age group served. Once again the Intrepid looked closely at several data sources, including participant performance and experience during the GOALS summers of 2009 and 2010, research on the expanded number of out-of-school-time programs, and availability of school-based opportunities for students in the middle school years. Evidence collected from such sources as the 2010 American Association of University Women (AAUW) publication *Why So Few? Women in Science, Technology, Engineering, and Mathematics* (Hill, Corbett, & Rose) and the 2011 U.S. Department of Commerce report *Women in STEM: A Gender Gap to Innovation* led to the shift to serve females entering the ninth and tenth grades. As more programs started for students at younger ages, a gap appeared for these girls transitioning into high school and an even wider gap for young women considering undergraduate or community college programs in STEM.

Expansions to the GOALS program since 2011 led to the current program structure, with components refined each year: a summer program with in-depth STEM exploration, workshops for alumnae during the school year, unpaid and paid internships, and part-time positions as junior educators open only to girls graduating from one of the museum's high school programs. Programs such as GOALS that transition participants from one experience to another as they age or develop specific skills are often referred to as pipeline programs. Feedback from GOALS girls each year articulates just such a need: a desire for continued programming after the summer portion ended, especially opportunities with pay. The Intrepid Museum's approach, which includes building career pipeline avenues, is supported by AAUW's (2013) call for increased opportunities engaging females in STEM during K–12 and higher education schooling and the organization's intimation that the community needs to collectively create environments that encourage girls to persist in STEM. The museum's pipeline structure for GOALS strives to do just this and work towards closing the gender gap through ongoing opportunities for alumnae as they navigate the next phases of their academic lives.

Revising Program Outcomes and Developing a Logic Model

Developing a formal evaluation strategy for the GOALS program began with reflections on assessment practices in past years and preparing a logic model with revised program outcomes. Logic models are planning guides that visually outline a program in a chart format. Learning outcomes are listed with outputs (activities, strategies, or products) that provide opportunities for the intended audience to achieve the outcome. For each outcome, the logic model details the data to be collected, data collection instrument(s), data collection procedure(s), and when the data will be collected.

The following three elements were considered in designing the new GOALS logic model:

- crafting specific outcomes that represent productive change for participants
- identifying specific outputs that allow for such change to be demonstrated
- developing evaluative questions that gauge the effectiveness of intended participant/program outcomes

GOALS Learning Outcomes and Outputs

Achievement of GOALS' learning outcomes varies among participants. Although outcomes are assessed for the group at large, each young woman's journey through the program is influenced by her personal characteristics, prior experiences and knowledge, and level of home and school support. To address these potentially differing responses, the logic model outlines various means of entry for its many outputs, and it defines success in a multifaceted way that considers increases in knowledge, changes in behavior and STEM interest, and demonstration of new skills. The learning outcomes for GOALS were revised to the following:

- To broaden student perspectives on STEM topics and careers
- To provide opportunities for students to practice problem-solving and critical-thinking skills
- To prepare students for subsequent STEM programming and educational pursuits, and ultimately STEM careers

Examples of outputs listed for outcome 3 include:

- *Summer Intensive:* goal-setting workshops; group definitions for professional and social behaviors; role-play of appropriate and inappropriate behaviors
- *Science Forum:* thematic workshops presented by Intrepid staff and guest facilitators from other non-profits, science institutions, STEM industries, and colleges
- *Navigator internship:* opportunities to attend professional networking programs and workshops throughout New York City; pairing with GOALS STEM advisory members as mentors

Although outputs can stand alone, in combination over time the experiences build the girls' confidence and encourage these young women to pursue STEM studies and careers. The outputs provide opportunities to address varying topics, behaviors, and skills; to support one another in learning and development; to explore STEM topics and career fields beyond textbooks and school subjects; and to network with like-minded young women in various environments and with professionals in a range of fields.

Evaluation Measures and Findings

Program evaluation questions are intended to provide insight into the program's influence on participants and to guide information to be collected during each key program component. The questions help ensure that program assessment measures align with program intent. Sample evaluation questions for each of the GOALS phases (Summer Intensive, science forums, and internship) are:

- How well do the outputs incorporate new knowledge and levels of challenging experiences for participants?
- What is the frequency of individual participation in the various science forums?
- How valuable are the effects on participants?
- How do we know we are reaching the audience we intended?
- What is the level of influence of the GOALS experience on participants' future school and career decisions?

As an out-of-school-time program, despite the strong academic focus, data collection instruments and procedures were designed to be fun and creative and administered as unobtrusively as possible so as not to resemble testing. Although inconspicuous procedures were not always practical and many involved asking participants to rate their experiences and respond to targeted questions, most methods incorporated observations, reviewing completed participant projects, crowdsourcing (posting questions on a poster throughout the program for ongoing, anonymous, yet public feedback/ideas/comments), journal writing, and tracking application and attendance data.

Many of the instruments are traditional in concept but modified for GOALS. A participant perception indicator (PPI), for example, is a tool designed to measure where an individual rates herself along a continuum in various areas. (See Figure 6.3.) The GOALS PPI strategically connects to the program's purpose, listing six target areas that include knowing about science topics other than those taught in school, being a confident problem solver and confident speaker, and personally meeting and speaking with STEM professionals. Participants in the Summer Intensive complete the PPI three times, so potential changes in response are tracked across the six weeks individually and as a group. Analysis of the PPIs provides details such as these: 82% of participants in the 2015 Summer Intensive reported that they felt confident in their STEM knowledge beyond that which they are taught in school, with 66% of the girls indicating improvement from the beginning to the end of the session; 82% reported confidence in their problem-solving skills, with 96% of the girls indicating improvement from the first day of the program.

Participant performance on worksheets completed during workshops provides formative feedback for making ongoing curricular adjustments, although these are not used for large-scale program evaluation because they are not summative measures. These daily worksheets establish a baseline for participants' understanding of topics and concepts, especially for instructional areas requiring revision to provide necessary scaffolding, whereas workbooks completed at the end of each two-week unit are analyzed for group mastery of key content.

Observation rubrics with performance indicators (see Figures 6.4 and 6.5) were created to assess each targeted program outcome by gathering quantitative and qualitative data on the program evaluation questions. Sample quantitative data include percent of participants making connections between science topics, asking questions and wanting more information, sharing ideas, and forming conclusions. Qualitative data involve documenting specific behaviors, comments made, and body language exhibited. Participants' journal entries and authentic reflection prompts provide qualitative details that give further insight into participant viewpoints, concept mastery, and perceptions. ("Journal entries" are kept in a journal given to the girls at the beginning of the program. Although encouraged, they are maintained organically by the girls over the six weeks. "Authentic reflection prompts" are individual prompts on separate pages provided to the girls at the conclusion of specific activities or sessions.)

INTREPID

GOALS Senior Navigator Interns

The following questions are designed to measure your perceptions regarding skills, confidence and experiences. Please read the statements carefully and circle the appropriate number and your yes/no response for each statement. Below are explanations for the numbers you circle when making your selections.

Please be honest. This helps us assess upcoming sessions to help you meet your goals.

1 - Beginning
I am not yet making significant progress towards this target

2 - Developing
I have not yet met this target, but I am making significant progress

3 - Accomplished
I have met this target

GOALS Sr. Navigator Targets	Beginning	Developing	Accomplished	Have you improved in this target area since entering the program?	What more can we do to help you in these areas?
1) Confidence expressing my ideas in front of a group.	1	2	3	Yes No	
2) Polished work-place skill sets including punctuality, professional dress, respecting my peers and supervisors, and following instructions.	1	2	3	Yes No	
3) Explaining scientific concepts to others, either in one-on-one conversations or in a group.	1	2	3	Yes No	
4) Involvement in STEM activities in school or other organization(s), outside of the Intrepid.	1	2	3	Yes No	

Figure 6.3 Participant perception indicator (PPI) for GOALS senior navigator interns.

Participant Observation Rubric--GOALS Science Forums

Date:	Observer:	Overall Program Theme for the Day:	Specific Activity:
Time of Activity:		Activity Presenter's / Facilitator's Name:	gallery program workshop
Length of Activity (in minutes or hours):			breakout activity large presentation
Approximate age(s) of participant(s) in activity:	As a multi-generational program: number of individuals in each of the following age breakdowns: Toddler/s ____ Children ____ Teen/s ____ Adult/s ____ Senior/s		Activity Location:

Level of engagement at beginning of activity	(low) 1	2	3	4	5 (high)	evidence for rating (behaviors, comments)?	
Level of engagement at end of activity	(low) 1	2	3	4	5 (high)	evidence for rating (behaviors, comments)?	

Attention and Relevance	****present STEM content and experiences that respond to interests, are engaging and develop interest****	
Participant(s) will:	Examples	Evidence = Descriptions/Notes
make observations	For large groups (ie: keynote speakers, presentations, programming in the Museum's spaces) ○ Majority of the group appears comfortable ○ Majority of the group is looking/focused ○ Majority of the participants make comments relating to prior knowledge	Provide notes to detail examples and provide evidence of what is observed (% of group, numbers of participants, body language, specific behaviors noticed, comments made, etc).
explore Museum artifacts and collections / engage in activities and demonstration	○ Majority of the participants make comments relating to current events or contemporary culture ○ Questions are asked ***	
build skills in social development, observation and communication	For smaller groups (ie: gallery programs, breakout activities, round robin format, etc) ○ Participants are looking/focused ○ Participants comment on what they see ○ Participants comment on what they do	
gain new knowledge and interest in range of disciplines in the STEM fields	○ Participants make comments relating to prior knowledge ○ Participants make comments relating to current events or contemporary culture	

INTREPID
SEA, AIR & SPACE MUSEUM

circle
one

Figure 6.4 Small- and large-group participant observation rubric for GOALS science forums.

INTREPID
Lab Table Challenges – Rubric

(Groups of 4 students are given 5–7 minutes to complete questions and/or tasks associated with Lab Table Challenges)

Assessor will observe the GROUP dynamics as the students move through the lab tables. Please stay with your associated table and complete the table below per each rotation:

ROTATION #1

| Group dynamics, communication and reasoning | Table #6: Group Names of students: _____ _____ _____ _____ | DIAGRAM
1. Using the large post-it paper, create an illustration of a Cartesian Diver. Draw and label the materials used in your image. | 1. # of students involved in completing the task _____

Are they speaking? Who is speaking? Who is engaged in the conversation? How are the students working together to complete the illustration? |
| | | SHORT ANSWER
2. Explain what happens to the "diver" when the bottle is squeezed and when it's released. How does this concept relate to what happens to the human body during a free dive water? | 2. # of students who complete task: _____

How do the students relate the concept to real-life situations? How do they share knowledge and ideas? |

Notes: _____

_____

Figure 6.5 Small-group participant observation rubric for GOALS science lab stations.

The crowdsourcing (the free-form and public response method) provides a safe space for the students to share their ideas, ask questions, and respond to one another. For example, student responses regarding a Science Forum that focused on the Hubble telescope documented that participants wanted to learn more about "how galaxies interact with each other in space," "how to be an engineer or mechanic," more about "coding to program the robotic arm" and "what other planets can we live on," and that they were surprised that "the robotic arm I made worked" and that "I am now more interested in space than I ever could imagine." Additionally, the observation rubrics completed by museum staff during varied activities documented girls' behaviors, interactions with one another, and engagement with the content, noting such things as making comparisons, asking for more materials, offering to help each other, commenting on what they saw and did, and trying more than one approach to solve a given problem or challenge. Students' authentic reflection prompts provide additional qualitative details. For example, a prompt given to Navigators about why they applied to be a Navigator Intern elicited responses such as, "GOALS was not enough, I had to come back," "there is something new every day," "I can actually make a difference and give back to my community," and "I have more chances to interact with STEAM [science, technology, engineering, art, and mathematics] and museum education."

Collectively, this information presents a more holistic view of the GOALS girls based on data that is intentionally measured and collected in relation to the program's stated outcomes. For example, responses from the 2014–2015 Navigator interns ($N = 15$) on their final survey reveals that all had an increased interest in STEM careers as a result of their GOALS experience, supported by such explanations as being "able to explore new topics and expand on the knowledge I already had through the GOALS program," "exploring the different branches of STEM," and "meeting so many inspirational women in STEM." All also reported having increased their grades in STEM subjects while participating in the GOALS program, with specific explanations including that GOALS "helped me learn not to procrastinate," "the Senior Navigator program helped me work with others my age, dealing with the same pressures, and who had similar goals to mine... this helped me keep motivated," "seeing how science is applicable and explaining it to a younger age group," and "having to present and talk to people on the museum floor helped me become more articulate which helped me when it came to presenting in class... I got higher grades on my presentations." In addition, all of the interns transitioning to college reported that GOALS inspired their decisions to focus on STEM studies, reporting that "before I was in this program I only had a faint idea about what field of study I belonged to, and meeting with professionals from these fields of study and hearing all about their careers motivated me to pursue subjects such as engineering and neuroscience" and "I have been very interested in medicine and engineering ever since we explored the two topics during my summer at Camp GOALS and further looked into during the College Fair and Be the Boss mentorship day. I hope to build a broad engineering foundation with Mechanical Engineering and pursue Biomedical Engineering in my graduate studies."

As previously stated, a variety of approaches were used for assessment purposes in order to avoid the feeling of being "tested." Although a gain in STEM content knowledge is certainly one of the intended GOALS program outcomes, there is a negative response to testing from many students in New York City, where high-stakes testing with little assessment value (the student gets no information on how to improve) and many negative consequences have been prevalent. For out-of-school-time programs supporting creativity, curiosity, and a willingness to try and fail, adjust, and try again that is necessary for innovation in STEM fields, it is essential not to reduce learning to the current test paradigm. In particular, programs for girls must incorporate assessments that move beyond measuring content knowledge gain into tracking potential increases in confidence and self-efficacy. Participant perception indicators are a particularly useful tool for this. Programs that aim to serve girls from backgrounds traditionally underrepresented in STEM and higher education also need to assess for success in increasing participants' awareness of STEM study and field options, interest areas, and ability to navigate situational cultural capital (verbal and non-verbal communication, behavioral expectations, etc.). Journaling, authentic reflection prompts, and observation rubrics are useful tools for tracking these.

GOALS FOR GIRLS: THE KEY INGREDIENTS

Creating a practical, comprehensive, and relatable STEM curriculum is challenging, but it is necessary to ensure participant engagement and development throughout the program. Current national mathematics and science standards and 21st century skills provide useful benchmarks for grade-level-appropriate program curriculum. Throughout the six-week program described in this chapter, the girls develop content knowledge and confidence by engaging in hands-on lessons grounded in the scientific method and the engineering design process.

The first step is getting girls who are underrepresented in STEM to enroll in the program. GOALS girls are not only underrepresented by gender, but also due to their racial/ethnic, linguistic, and socioeconomic backgrounds. For any program focused on supporting STEM talent from underrepresented groups, the key is encouragement. Intrepid Sea, Air & Space Museum education staff members encourage program participants in every way possible. We encourage them to apply to the program through outreach not only to schools but also to community centers and afterschool programs. We make it easier to complete the application by offering digital and paper options, and when we observe partially completed applications, we reach out to those individuals to encourage them to complete the final pieces. We encourage the girls to speak up during the program, as well as to ensure that their peers also have a voice. We encourage families to be involved in the program and participants to pursue other opportunities we recommend. For any program aiming to make a larger impact on the pipeline of girls and other underrepresented groups in STEM, these extra efforts of encouragement must be co-equal in importance with the STEM content itself.

REFERENCES

American Association of University Women. (2013). *Quick facts: STEM.* Retrieved from http://www.aauw.org/files/2013/09/quick-facts-on-STEM.pdf

Cech, S. (2007). Much of learning gap blamed on summer. *Education Week, 26*(43), 5, 15. Retrieved from http://www.edweek.org/ew/articles/2007/07/18/43summer .h26.html

ExpandED Schools. (n.d.). *The 6000 hour learning gap.* Retrieved from http://www. expandedschools.org/sites/default/files/tasc_6000-hours-infographic.pdf

Hill, C., Corbett, C., & Rose, A. (2010). *Why so few?: Women in science, technology, engineering, and mathematics.* Washington, DC: American Association of University Women.

Institute of Museum and Library Services. (2013). *Growing young minds: How museums and libraries create lifelong learners.* Washington, DC: Author.

McCreedy, D., & Dierking, L. (2013). *Cascading influences: Long-term impacts of informal STEM experiences for girls.* Philadelphia, PA: The Franklin Institute.

Modi, K., Schoenberg, J., & Salmond, K. (2012). *Generation STEM: What girls say about science, technology, engineering, and math.* New York, NY: Girl Scout Research Institute. Retrieved from http://www.girlscouts.org/content/dam/ girlscouts-gsusa/forms-and-documents/about-girl-scouts/research/generation_stem_full_report.pdf

Munley, M. E., & Rossiter, C. (2013). *Girls, equity and STEM in informal learning settings: A review of literature.* The Girls RISE National Museum Network. Retrieved from http://stem.arizona.edu/sites/stem.arizona.edu/files/SAVI%20Lit%20 Review%20Oct%202013%20Girls%20STEM.pdf

Partnership for 21st Century Skills. (n.d.) *Framework for 21st century learning.* Retrieved from http://www.p21.org/our-work/p21-framework

U.S. Department of Commerce. (2011). *Women in STEM: A gender gap to innovation.* Retrieved from http://www.esa.doc.gov/sites/default/files/womeninstem agaptoinnovation8311.pdf

White House Office of Science and Technology Policy. (2014, March). *Preparing Americans with 21st century skills.* Retrieved from https://www.whitehouse.gov/sites/default/files/microsites/ostp/Fy%202015%20STEM%20ed.pdf

CHAPTER 7

ALL GIRLS/ALL MATH SUMMER CAMP

Wendy Hines and Lindsay Augustyn

ABSTRACT

All Girls/All Math is a seven-day summer camp for high school girls from across the United States who are interested in mathematics. This residential camp held at the University of Nebraska–Lincoln has been conducted annually since 1997. Acceptance is highly competitive, with 50 to 60 girls (25 to 30 for each of two camps) selected to participate each year based on availability of funding. At the camp, the girls take a course in cryptography and number theory, working their way up to RSA cryptography. This course meets three hours a day for five days. The other half of the camp day is spent in self-contained mini-courses—one three-hour mini-course per day on varying topics. Program guests include a career panel and a visit from a National Security Agency (NSA) mathematician. Recreational activities, such as laser tag, rock climbing, a movie, and a picnic are held in the evenings to help build camaraderie. Participant evaluations of the program indicate that the camp is very successful. The girls rate their experience highly and express enhanced understanding of the uses and types of mathematics, as well as greater interest in pursuing mathematics careers.

Out-of-School-Time STEM Programs for Females, pages 151–167
Copyright © 2017 by Information Age Publishing

ALL GIRLS/ALL MATH SUMMER CAMP

All Girls/All Math is a seven-day residential summer camp for high school girls interested in mathematics held at the University of Nebraska–Lincoln. Each summer the program includes two identical camps of 25 to 30 girls each. The camp, which began in 1997, was developed in response to research findings that girls do better in an all-female environment and with the intent to encourage girls to go into STEM (science, technology, engineering, mathematics) fields. The program creators, Wendy Hines and Judy Walker, originally based the camp on their own experiences and on what they believed would have helped them. Since then, the camp has gone through some changes based on feedback provided on end-of-camp evaluations completed by program participants. For example, originally the campers took two courses, one in cryptography and number theory and one in chaos, but the girls found it hard to keep up with two challenging courses, and they wanted to be exposed to more mathematics topics. Therefore, one course was replaced with daily self-contained mini-courses. One advantage of the mini-courses is that the girls experience a broader spectrum of mathematics.

The camp has been listed in Kaplan's *Yale Summer News Guide to Summer Programs*; the book *The Ultimate Guide to Summer Program for Teens*, which lists the editors' choice of the top 200 summer programs nationwide; the guide *Discover Your Summer* produced by the Chicago Math and Science Initiative; and many other publications. It is also listed on the American Mathematical Society's website in addition to having its own website, which may be found at http://www.math.unl.edu/programs/agam. Many girls and their parents find the All Girls/All Math program while doing web searches.

During the camp, the girls take a five-day course in cryptography and number theory with the goal being to introduce them to RSA cryptography. This course, which meets three hours each morning, combines lecture and group work. Each afternoon, the girls have a self-contained mini-course on a different subject. Sample mini-course topics have included aerodynamics, graph theory, statistics, population genetics, knot theory, fractals, mathematics within the game of Set, Boolean networks in biology, and bioinformatics. The mini-courses expose the girls to various mathematical ideas. A career panel is incorporated into the camp to demonstrate mathematics-related occupations. A National Security Agency (NSA) mathematician visits the camp to talk about working at the NSA and to show mathematics related to her work and to the cryptography course.

The girls also engage in recreational activities that are mainly designed to build camaraderies. Sample lunchtime activities include going to an ice cream shop, having a campus tour, and playing sand volleyball. Evening activities include laser tag, rock climbing, a picnic, and going to a movie.

RELATED LITERATURE

Throughout much of the world, boys continue to outscore girls on standardized mathematics tests. For example, in most of the 57 countries that participated in the Programme for International Student Assessment (PISA) in 2006, boys' performance was significantly higher than girls on the mathematics scale (Wiest, 2008). Women earn 41% of bachelor's degrees in mathematics and only 25% of mathematics PhD's, despite the fact that 52.2% of all PhD's go to women (Perry, 2013). Research has consistently shown that girls have lower mathematics and science confidence and self-concepts than boys (e.g., Ross, Scott, & Bruce, 2012; Tariq, Qualter, Roberts, Appleby, & Barnes, 2013) and that confidence predicts mathematics achievement (Stankov, Lee, Luo, & Hogan, 2012).

Some research has shown benefits for girls who take mathematics in a single-gender environment (McFarland, Benson, & MacFarland, 2011; Picho & Stephens, 2012; Shapka, 2009; Tully & Jacobs, 2010). For example, McFarland et al. (2011) conducted a study at a Midwestern elementary school in which they separated a class of fifth-grade girls and boys into different classrooms and compared same-gender and mixed-gender student achievement after one year. Mathematics scores for females in the same-gender classroom were higher than scores for females in the mixed-gender classrooms. Additionally, females in the girls-only classroom scored higher in mathematics than males in both boys-only and mixed-gender classrooms.

Girls-only out-of-school-time programs, such as summer camps, have been shown to favorably influence a number of mathematics-related outcomes for participants, such as greater STEM participation (McCreedy & Dierking, 2013; Wiest, 2010). For example, Wiest (2008) investigated the influence of a one-week mathematics and technology camp for middle school girls through entrance, exit, and fall follow-up surveys designed to assess selected mathematics-related attitudes, as well as perspectives on program effectiveness and quality. Findings indicate successful program outcomes. Participants showed statistically significant improvements in attitude and reported increased knowledge and skills, improved grades, and greater participation in mathematics and technology. At the high school level, Chacon and Soto-Johnson (2003) studied 36 girls who attended a mathematics summer camp. The girls completed pre/post surveys, kept a daily reflection log of learning, and later completed a follow-up survey. The researchers found that attending the one-week summer camp resulted in statistically significant improvements in participant attitudes, confidence levels, persistence, and perceptions of the value of peer collaboration.

Role models have also been shown to favorably influence females in mathematics. Krings (2014) found a correlation between the number of women who earned master's degrees in mathematics and the percentage of women at that university, whether faculty, administrators, or fellow students. This might mean that high school girls are more likely to pursue mathematics if they have more female role models and meet more female mathematics students like themselves.

PROGRAM DESCRIPTION

The All Girls/All Math summer camp starts on a Sunday afternoon, during which participants check in with camp coordinators, spend time getting to know each other, and play games. A University of Nebraska–Lincoln (UNL) faculty member in mathematics delivers a welcome address before dinner. Classes start on Monday morning and end on Friday afternoon. A competitive problem-solving contest takes place Saturday morning, and camp ends that day at noon. The girls spend the final morning playing games, working on mathematics problems, and saying goodbye to each other. The sections that follow describe the camp curriculum, the application procedure, camp tuition, program personnel, and specific camp operations.

Course Descriptions

Cryptography

The cryptography course is the main course conducted across the camp week. It is held for three hours each morning. Approximately one hour of each class is devoted to lecture. During the remaining time, the girls work together on problem sets in groups of two or three. The course begins with an introduction to cryptography through history. The Spartan Scytale is explored as an example of a transposition cipher, followed by the Caesar cipher and, more generally, substitution ciphers. A discussion is held about the statistical approach to decoding substitution ciphers, which the girls practice with the help of a computer to count frequencies and to implement the girls' ideas about what the substitutions should be. This helps illustrate the idea that these ciphers are not very secure and sets the stage for a more mathematical approach. The girls are introduced to modular arithmetic and the idea of the greatest common divisor (GCD) of two integers, connecting the GCD with solving linear congruence equations. Next they are taught the Extended Euclidean Algorithm so they can compute

GCD's and solve equations. All of this is first done by hand. Once the girls understand the concepts and procedures, they learn how to use Maple technical computing software to do much of the computations for them. From there, the girls engage in a discussion about exponentiation modulo n and the idea of "fast exponentiation." Again, this is first done by hand so the girls understand it is feasible before they are shown how to do it using Maple. The group then moves on to Euler's theorem, which is the basis for RSA public key cryptography. Finally, a discussion of public key cryptography ensues. At this point, the girls have all the background knowledge they need to understand the mechanics of the RSA system, and they get very excited by the idea that all of this mathematics has such a practical application. After a discussion of how the RSA system works, the girls do a task that shows encryption is not secure if the primes chosen when developing keys are too small. They also see, via use of Maple, that while finding relatively large primes is "easy," factoring a product of two large primes is "hard." (For example, Maple's nextprime function can find 20-digit primes instantaneously, but Maple's ifactor function cannot factor a 40-digit number formed by multiplying two of these primes together.) Finally, each girl sets up her own public and private key system using 20-digit primes, and everyone puts their public keys in a central location on the computer. The girls have great fun using the public keys to send encrypted email to one another. The course ends with a treasure hunt where the clues are given in code. Each clue is given in a different code, including RSA. The girls have to determine which code is being used and then crack the code. This is a favorite activity within this course.

Aerodynamics

The focus of this mini-course (a stand-alone three-hour course held in the afternoon) is mainly on how lift is generated. With staff support, the girls derive Bernoulli's law relating change in pressure (p) with change in velocity of the flow (v):

$$\nabla p = -\rho v \nabla v$$

where ρ is the density of the medium. Experiments are conducted for the girls to see how lift is generated by a velocity differential (e.g., a wind tunnel is created by running a vacuum cleaner in reverse). Methods for creating a velocity differential are then discussed, such as building an airfoil that can slide up and down on a string and putting it in our wind tunnel. Finally, other mechanisms to generate lift are considered, including incidence force and the role of turbulent flow.

Knot Theory

The girls study knotted circles in three-dimensional space. Two knots are considered the same if one can be bent and twisted to look like the other without cutting the circle or letting it pass through itself (a maneuver that is possible in mathematics, although not with a real circle of rope). The main motivating problem in the theory is to find effective methods to decide whether two knots are really the same. The girls explore this problem and possible solutions. In some sense it turns out to be "easier" to determine that two knots are different than it is to find they are the same. This seems to be true even when an individual knows that one is the "simplest" of all knots: the unknot.

Mathematics within the game of Set.   Set is a card game that is based (in disguised form) on the affine four-dimensional geometry over $\mathbb{Z}_3$. Each card shows a collection of colored, shaded, and shaped symbols. The points of the geometry are the cards, and the object of the game is to recognize a "set"—three cards that form in affine line in $\mathbb{Z}_3^4$. Connections with topology are discussed (the two-dimensional version of the game is simply Tic-Tac-Toe on a torus) and with Szemeredi's theorem concerning the existence of arithmetic progressions in dense subsets of the integers (an affine line is also, of course, an arithmetic progression).

Fractal geometry.   This mini-course begins with a discussion of linear transformations in the plane—translations, reflections, rotations, and shears—and what these transformations do to plane figures. Unions of transformations are then considered. That is, the girls must determine when f, g, and h are transformations and A is a set in the plane, what image does the following give us?

$$F(A) := f(A) \cup g(A) \cup h(A)$$

The girls learn that applying the function F iteratively yields, in the limit, a fractal, and they determine the transformations that create the Sierpinski Triangle and other simple, but famous, fractals. The next thing discussed is the Collage Theorem, which indicates how to determine the transformations based on the desired final image. The girls are given a fractal image for which they must determine the transformations used to create it and then test their transformations using a staff-written program. The course concludes with the girls viewing some of Kent Musgrave's fractal art and discussing fractals as a tool for image compression.

Boolean networks in biology.   The logical operators *and, or, not,* and Boolean values are introduced, along with a brief description of the process of transcription of genes and translation of mRNA to produce proteins. Next, the girls see how the qualitative properties of gene regulation can be represented by Boolean functions and that each Boolean function can represent a specific biological property (e.g., *and* = synergistic regulation, *not* = inhibition). The girls do computations with Boolean functions and show how they describe biological processes. Sharing a paper published in a professional journal shows how Boolean networks are being used in current research. The course then shifts from Boolean functions to Boolean networks that are used to describe a biological system. Properties of Boolean networks (trajectories, steady states, and dynamics) are studied with attention to how they correspond to the dynamic behavior of biological systems. Software is used to compute the dynamics of Boolean networks and show the limitations of computer software, which motivates the theoretical study of Boolean networks. Analyzing Boolean networks allows the girls to identify patterns and note that the patterns seem to always occur. We briefly present a paper published in a professional journal that shows the patterns have been proven to always occur.

Application Procedure

Every year, camp fliers are sent to all high schools in Nebraska. In the past, a list of girls who performed well on the American Mathematics Competitions test was also obtained from the Mathematical Association of America, and we sent letters to those girls. More recently, the camp sends letters to schools that had several students perform well on the American Mathematics Competitions test. Finally, the program is listed on the American Mathematical Society's and our own program website (http://www.math. un.edu/programs/agam), which is often found by several campers doing web searches.

Applicants must submit their grade transcripts, a letter of recommendation, and an essay addressing why they want to attend the camp. Acceptance is highly competitive. Transcripts are checked for sufficient mathematics background (i.e., algebra and geometry), and all selected girls have typically received grades equal to or above B+ in mathematics courses. We read applicants' recommendation letters and essays to learn about their interests; how well they have done in their studies, which is not limited to their mathematics courses; their personality; and other such factors. Typically, 25 to 30 girls are accepted into each of the two camps from about 70 to 80 applicants from around the country. Applicants have come from every U.S. state, except Hawaii, and some from other countries, such as Belgium and South Korea.

Special consideration is given to applicants entering the tenth and eleventh grades based on an assumption that a girl entering the twelfth grade is more apt to have already decided on a college major. Greater preference is also given to girls who have not yet taken calculus, because girls who take calculus tend to be more likely to enter a science-related career (Ma & Johnson, 2008). Girls must have taken geometry to be accepted into the camp because we believe geometry gives them the level of mathematical maturity that is needed for the program.

All other things being equal, we choose girls from rural towns and inner cities, reasoning that these girls are the most isolated mathematically and can benefit most from a program like ours. A few inner city girls have participated in the program despite difficulty attracting these girls. Because the camp is located in Lincoln, Nebraska, it attracts many rural students due to their comfort level traveling to this small city and to Lincoln's reputation as a friendly college town. This is important because rural girls tend not to have as many opportunities to attend a summer camp in mathematics as do girls in urban areas.

Tuition and Funding

Until recently, tuition has been $125 for Nebraska girls and $250 for out-of-state girls. We offered $125 tuition scholarships and $500 travel scholarships to make camp participation reasonably accessible to interested girls. We recently shifted to a voluntary tiered pricing system, allowing families to choose the price that works best for them. Out-of-state families may choose either $500 or $350, and Nebraska families may choose from $500, $350, or $175.

The actual per-student cost for the camp, including the cost of providing instruction, housing, meals, and social activities, is a bit more than $1,000. Because our tuition is, at most, half of this true cost, our annual funding request tends to be high. Our most recent total budget was about $56,000. This covers such costs as: camper room and board; undergraduate coordinator room and board; recreation center fees for the campers; salary, room and board, and travel for instructors; travel and room and board for the visiting mathematician from the National Security Agency; salary and room and board for a graduate assistant; and postage and printing. UNL pays the salary of the undergraduate student staff member (described below in "Personnel").

We currently have three types of scholarships: one based on grades earned in mathematics courses (awarded to one girl), one based on grades from among those girls who applied for tuition and travel scholarships (awarded to one girl), and the third based on financial need (awarded to any girls who need support to attend the camp).

Proposals for financial support are also sent to a number of potential funding sources, such as the AMS (American Mathematical Society) Epsilon Fund, the National Security Agency's STEM Education Partnership

Program, and the Mathematical Association of America/Tensor Foundation grant program. The All Girls/All Math Summer Camp is one of the few programs regularly supported by the AMS Epsilon program.

Personnel

Each camper group has one instructor. The instructors are women who received their PhD in mathematics from UNL. A female graduate assistant and a female undergraduate teaching assistant serve as chaperones in the dormitory, take the girls to evening activities, and supervise them. The undergraduate teaching assistant also serves as the undergraduate coordinator before the camp, doing the bulk of the work to organize the camp and communicate with participants before the camp. The camp director performs such tasks as writing proposals to attain program funding, choosing campers from the applicant pool, and choosing the undergraduate coordinator (see others in "Program Operations" below). Mini-course teachers may be female or male. The mini-course teachers volunteer their time, although they are each given a $100 honorarium. Volunteers drive the girls to activities in large university vans and pick up and drop off girls at the airport. The five people on the career panel also volunteer their time. This panel is typically all female, but it occasionally includes some male panelists.

A faculty member in the UNL Department of Mathematics gives a welcome speech to the campers. She or he also proposes two or three challenging mathematics problems to the girls and offers a monetary award ($3 to $10, depending on problem difficulty) to the campers who solve each problem first. The girls form their own groups, which are usually different for each problem. The activity works well for team building, as the girls discover they need to form bigger groups for the harder problems with the larger monetary award. The girls split the award money evenly between the winning group members for each problem. Finally, the accountant for UNL's Center for Science, Mathematics and Computer Education (CSMCE) handles the budget for this program, and the events coordinator supervises the undergraduate coordinator, ensures the event's compliance with university regulations (youth safety policy), and is responsible for any associated contracts (housing, vans, food).

Program Operations

The first thing that the CSMCE events coordinator does is to reserve camper dormitory rooms. This must be done mid-fall, as rooms fill up fast. The CSMCE revises the program website in the fall so that dates are accurate and creates all advertising materials. The CSMCE accountant develops a budget, and the camp director writes proposals to submit to several potential funding sources. The undergraduate coordinator sends advertisements about the camp to all Nebraska high schools and sends letters to girls who have performed well on the AMC test, inviting them to apply to the camp. The camp director asks the NSA to provide names of volunteers to serve as guest speakers during the camp, and she finds instructors, mini-course instructors, and volunteer drivers. The CSMCE events coordinator arranges travel and room and board for the instructors and NSA visitors. The CSMCE also serves as a resource for the mini-course instructors by providing needed materials and support (for example, materials to make airfoils in the aerodynamics mini-course and Frisbees for the aerodynamics mini-course).

The undergraduate coordinator assembles camper applications as they are submitted online. She compiles the documents electronically into one application file for each applicant and gives them to the director, who then chooses the campers based on the criteria noted earlier. The undergraduate coordinator also places campers in groups of two classes for each camp. First, we try to satisfy all girls' preference for which week they would like to attend. Our goal is to make sure that the two camps have almost the same number of girls, and each camp has two classes with about the same number of girls. An attempt is made to group girls based on experience. We divide each camp into two codes classes, one comprised of students with a stronger mathematics background and one with those who have a relatively weaker mathematics background. The groups are chosen so that a girl who is first in her class of 20 in small-town Nebraska is not in the same group as a girl who is first in her class at the Illinois Math and Science Academy, for example. The group with less experience is taught by an instructor who has previously taught at the camp.

The undergraduate coordinator handles all communication with campers. Once the campers are chosen, the undergraduate coordinator sends acceptance and rejection letters to all applicants and decides which campers to place in which camp based on campers' level of mathematics experience and preference for camp date (two identical camps are run two different weeks). The undergraduate coordinator sends letters to the accepted campers telling them what to bring and what to expect, and she sends them a form to complete that provides helpful information for assigning roommates, such as whether they are a night or morning person. She also sends appropriate permission slips. She gets flight information for the campers who fly into Lincoln and coordinates airport pick-ups and drop-offs with the CSMCE. The undergraduate coordinator works with the events coordinator to choose and create a schedule for evening activities, which must be approved by the director. The CSMCE events coordinator and accountant assume responsibility for ensuring that all program staff who will have contact with the campers have a background check and sign a safety agreement form as per university policy.

PROGRAM EVALUATION

The girls complete an evaluation form (see Appendix A) on the last day of the camp. The evaluation form consists of 30 statements that the girls rate from 1 (strongly disagree) to 5 (strongly agree). The statements are followed by three open-ended, short-answer questions. Table 7.1 shows some of the noteworthy results from the rated statements on the 2014 camp evaluations, completed by 47 girls.

Mean ratings for #3 in Table 7.1 are approximately the same each year. The high standard deviation for that statement (compared to most other statements) indicates that while the average was a little over 3 on the 1 to 5 scale, there was a larger spread in the responses, with many students rating this item higher (4–5) or lower (1–2). The larger standard deviation tells us the comfort level with discussion had significant variation, with some girls being much more comfortable than others as compared to their comfort level with having mathematical discussions at their school. In 2014, we wanted to survey past participants who were now in college or beyond. We sent surveys to a subset of participants who attended the camp during the years 1997 to 2010 (via email if we had a valid email address, or via a postcard with a link to the survey if we had a valid mailing address); 43 of the 65 women responded. When statement #3 was rated again in this survey of former campers, the mean rating was 4.10. It appears that after the girls returned to school, they

TABLE 7.1 Selected Mean Ratings and Standard Deviations for the Exit Survey Completed by Campers

Statement	SD	M
1. My camp experience has increased my interest in studying mathematics.	4.67	.49
2. The female-focused atmosphere was a valuable aspect of the camp.	4.70	.59
3. At the camp, I was more comfortable participating in the discussions than I usually am at school.	3.38	1.11
4. I learned new information and skills in the sessions on codes.	4.95	.20
5. My camp experience was fun.	4.76	.43
6. I learned more about mathematics-related careers that are available to me.	4.33	.73
7. When I return home, I plan to follow up on what I've learned.	4.10	1.00

Note: Ratings ranged from 1 (strongly disagree) to 5 (strongly agree).

became especially aware that they were indeed more comfortable participating in discussions than they had been previously.

Statement #7 had the second largest standard deviation. While the majority of girls indicated that they did plan to implement some or all of what they learned when they returned home, a smaller group of girls rated this item low, perhaps not seeing where they could apply what they learned at camp back at home. Some girls with fairly traditional high school mathematics programs might not see where knowledge of knot theory and RSA cryptography might fit.

Below are sample responses to the open-ended question, "What were the best parts of this camp experience for you?" The comments represent the main themes appearing in the responses:

- Honestly, I loved getting to know all of the girls and being able to totally nerd out. I enjoyed learning all about the codes and solving the problems. I loved pretty much everything.
- I loved learning about cryptology and codes the best.
- I was able to meet other people with similar interests and challenge myself in mathematical areas, also, I like learning about cryptography and how it could apply to everyday life.
- I really enjoyed the codes class—it was a lot more stimulating and interesting than my class at school. The girls were really nice, which was great.
- Coding class was awesome, just very difficult. Scavenger hunt, bonding with other girls, laser tag, ice cream, free time.

The girls clearly enjoyed meeting girls like themselves and liked the codes course (called the cryptography and number theory course earlier). When asked what would have improved their camp experience, the girls often mentioned that the codes course was a little too long and that they would have enjoyed more freedom. The girls had less freedom in the year in which this survey took place than in previous years because of UNL's increased safety measures. A chaperone had to accompany them everywhere. We now hire a third chaperone so that the girls may do more things away from the main group.

The final open-ended question was: "Take a few moments to reflect on your experience at this camp and how this has affected your interests, your future goals, and your attitude toward mathematics. How have these things changed in the last week?" All camper responses, such as the following examples, were positive.

- I am much more interested in math and feel more motivated to pursue a career in math. I now know how wide the variety of math subjects is and finally learning coding was very interesting and rewarding. I will definitely look into job shadowing, future math/ science camps, and will recommend this camp to others! Thanks for an amazing week.

- I definitely feel empowered as a female that I can be a successful mathematician, the career panel was great in both showing me that fantastic females are in math-related fields as well as to help me learn more about careers that I didn't know about. This camp definitely solidified my love and passion for math.

- I knew I wanted to pursue something in STEM, but this camp made me realize it might be pure math.

- I took an interest in college math, because this stuff is nothing I've ever seen before. I also met a bunch of amazingly talented people, who shared my interest in solving problems and not giving up. Even at a math school, I had not encountered this. Now that I know there are more of me, I feel more driven to excel in math, even when the kids in my grade do not. I realized that math was a lot of fun.

- I've never considered majoring in math but now I definitely think it would be a good choice to do so.

- Honestly, after going to this camp, I am so much more confident in my mathematical abilities and my potential in math. I had no idea what I wanted to do with my future regarding a career and now I do. I want to do something with math. This camp opened my eyes to numerous possibilities and I am so grateful to everyone who made this camp happen because you are really helping to shape futures and change lives and outlooks on life. Thank you so much! I'll never forget this camp.

- I'm definitely more interested in studying math in college (maybe even majoring/minoring?)

- I think I really want to go into a math career! I've always enjoyed math but it's sort of boring. Being challenged for the first time in math made me realize how much I love it. I might want to double major in math and something else (journalism or poly science) and then pursue a higher degree in a math field.

The girls' responses to this final question show that their understanding of the types and uses of mathematics had expanded. They also expressed having developed greater interest in pursuing a mathematics degree. Taken together, nearly all responses fit into one of these two categories, in which girls were enthusiastic about how much they had learned and in their plans to pursue a mathematics (or STEM) major in college and/or their career.

CLOSING COMMENTS

In today's world in which high school students might eventually take jobs that do not currently exist, given the advances in technology, having students with strong problem-solving skills is essential. Participants in All Girls/All Math gain exposure to problem solving in different situations and learn to utilize different skills in STEM fields, thus helping to prepare them for future STEM careers. Camps that focus on females, such as All Girls/All Math, fill a societal need, given the preponderance of evidence that females do not participate in mathematics (or STEM) fields at the same rate as males. Giving students exposure to the beauty of mathematics and confidence in their problem-solving skills can help to address females' low participation in mathematics fields.

In order for another site to replicate the All Girls/All Math camp experience for girls, some challenges to address include: securing a sufficient number of staff members to coordinate the advertising and logistics of the event; acquiring adequate funding to make the camp affordable; having a sufficiently large pool of female mathematicians to recruit as instructors; and planning team-building activities (such as rock-climbing; considerations include any necessary transportation). In recent years, as the university became more safety conscious about youth activities, ensuring the camp activities and camp volunteers were all in compliance with the safety policy added a level of bureaucracy to the camp logistics. While mathematicians often conceive of and direct camps like All Girls/All Math, staff support is a crucial component to ensure everything runs smoothly and observes university policies.

Based on the camp evaluations, it is apparent that the All Girls/All Math program has successfully met its goals. By the end of the camp the girls were more interested in pursuing a mathematics degree than they were before the camp. The experience of meeting other girls like themselves was very positive and valuable. The girls enjoyed being challenged by some mathematics that they would never see in high school. They also rated the same-gender nature of the camp highly, lending support to other studies that have found positive results for girls-only STEM learning environments (e.g., McFarland et al., 2011; Picho & Stephens, 2012; Shapka, 2009; Tully & Jacobs, 2010). The fact that the girls reported a highly positive camp experience is crucial, because dispositions can influence girls' mathematics experiences and continued participation in the subject (e.g., Stankov et al., 2012). As one girl stated, "This was the best week of my life!"

APPENDIX A

Evaluation form program participants completed at the end of the camp week. *Note:* On the original form the column choice headings were fully written out as follows: *SD* = Strongly Disagree; *D* = Disagree; *N* = Neither Agree Nor Disagree; *A* = Agree; *SA* = Strongly Agree.

All Girls/All Math 2014 Evaluation Form

Please circle the appropriate number that best describes your response to the statements below.

	SD	D	N	A	SA
I had an opportunity to meet girls with math interests similar to mine.	1	2	3	4	5
My camp experience has increased my interest in studying mathematics.	1	2	3	4	5
The female-focused atmosphere was a valuable aspect of the camp.	1	2	3	4	5
I plan to stay in touch with some of the girls that I met this week.	1	2	3	4	5
At the camp, I was more comfortable participating in the discussions than I usually am at school.	1	2	3	4	5
I learned new information and skills in the sessions on mathematical biology.	1	2	3	4	5
I learned new information and skills in the sessions on graph theory.	1	2	3	4	5
I learned new information and skills in the sessions on aerodynamics.	1	2	3	4	5
I learned new information and skills in the sessions on the mathematics behind the game of Set.	1	2	3	4	5
I learned new information and skills in the sessions on codes.	1	2	3	4	5
I view women mathematicians that I met as role models.	1	2	3	4	5
My camp experience was fun.	1	2	3	4	5
The amount of work required was manageable for me.	1	2	3	4	5
I felt like I belonged at this camp.	1	2	3	4	5
I would recommend this camp to my friends.	1	2	3	4	5
This was my first opportunity to meet women mathematicians.	1	2	3	4	5
I felt lost during the codes class.	1	2	3	4	5

I learned more about mathematics-related careers that are available to me.	1	2	3	4	5
I thought that working the problems increased my understanding of the course material in the codes class.	1	2	3	4	5
The format of the lectures and problem sessions was a good way for me to learn.	1	2	3	4	5
I plan on sharing what I've learned with my classmates at school.	1	2	3	4	5
I enjoyed the mathematical topics that were covered in the math biology course.	1	2	3	4	5
I enjoyed the mathematical topics that were covered in the graph theory course.	1	2	3	4	5
I enjoyed the mathematical topics that were covered in the aerodynamics course.	1	2	3	4	5
I enjoyed the mathematical topics that were covered in the mathematics behind the game of Set.	1	2	3	4	5
I enjoyed the mathematical topics that were covered in the codes class.	1	2	3	4	5
I enjoyed having the activities to participate in. My favorite was ____.	1	2	3	4	5
I did not enjoy this camp.	1	2	3	4	5
This camp was a great experience for me.	1	2	3	4	5
When I return home, I plan to follow up on what I've learned. Please describe:	1	2	3	4	5

Please respond to each of the following questions.

1. What were the best parts of this camp experience for you?

2. What could have improved your camp experience?

3. Take a few moments to reflect on your experience at this camp and how this has affected your interests, your future goals, and your attitude toward mathematics. How have these things changed in the last week?

We enjoyed getting to know each of you!
THANK YOU for an awesome week!

REFERENCES

Chacon, P., & Soto-Johnson, H. (2003). Encouraging young women to stay in the mathematics pipeline: Mathematics camps for young women. *School Science and Mathematics, 103*(6), 274–284.

Krings, M. (2014). *Study explores where high number of women earn STEM degrees.* Retrieved from http://phys.org/news/2014-05-explores-high-women-stem-degrees.html

Ma, X., & Johnson, W. (2008). Mathematics as the critical filter: Curricular effects on gendered career choices. In H. M. G. Watt & J. S. Eccles (Eds.), *Gender and occupational outcomes: Longitudinal assessment of individual, social, and cultural influences* (pp. 55–83). Washington, DC: American Psychological Association.

McCreedy, D., & Dierking, L. D. (2013). *Cascading influences: Long-term impacts of informal STEM experiences for girls.* Philadelphia, PA: The Franklin Institute. Retrieved from https://www.fi.edu/sites/default/files/cascading-influences.pdf

McFarland, M., Benson, A., M., & MacFarland, B. (2011). Comparing achievement scores of students in gender specific classrooms with students in traditional classrooms. *International Journal of Psychology*, no. 8, 99–114.

Perry, M. J. (2013) *Women earned majority of doctoral degrees in 2012 for fourth straight year, and outnumber men in grad school 141 to 100.* Retrieved from http://www.aei-ideas.org/2013/09/women-earned-majority-of-doctoral-degrees-in-2012-for-4th-straight-year-and-outnumber-men-in-grad-school-141-to-100/

Picho, K., & Stephens, J. M. (2012). Culture, context and stereotype threat: A comparative analysis of young Ugandan women in coed and single-sex schools. *Journal of Educational Research, 105*(1), 52–63.

Ross, J. A., Scott, G., & Bruce, C.D. (2012). The gender confidence gap in fractions knowledge: Gender differences in student belief-achievement relationships. *School Science and Mathematics, 112*(5), 278–288.

Shapka, J. (2009). Trajectories of math achievement and perceived math competence over high school and postsecondary education: Effects of an all-girl curriculum in high school. *Educational Research and Evaluation, 15*(6), 527–541.

Stankov, L., Lee, J., Luo, W., & Hogan, D. J. (2012). Confidence: A better predictor of academic achievement than self-efficacy, self-concept and anxiety? *Learning and Individual Differences, 22*(6), 747–758.

Tariq, V. N., Qualter, P., Roberts, S., Appleby, Y., & Barnes, L. (2013). Mathematical literacy in undergraduates: Role of gender, emotional intelligence and emotional self-efficacy. *International Journal of Mathematical Education in Science and Technology, 44*(8), 1143–1159.

Tully, D., & Jacobs, B. (2010). Effects of single-gender mathematics classrooms on self-perception of mathematical ability and post secondary engineering paths: An Australian case study. *European Journal of Engineering Education, 35*(4), 455–467.

Wiest, L. (2008). Conducting a mathematics camp for girls and other mathematics enthusiasts. *The Australian Mathematics Teacher, 64*(4), 17–24.

Wiest, L. R. (2010). Out-of-school-time (OST) programs as mathematics support for females. In H. J. Forgasz, J. R. Becker, K.-H. Lee, & O. B. Steinthorsdottir (Eds.), *International perspectives on gender and mathematics education* (pp. 55–86). Charlotte, NC: Information Age.

CHAPTER 8

CONCLUDING THOUGHTS

What We Have Learned and What We Still Want to Know

Lynda R. Wiest, Jafeth E. Sanchez, and Heather Glynn Crawford-Ferre

The professional literature reported in the chapters of this book converge on two main concerns for females in relation to STEM. One is the underrepresentation of females in STEM when these subjects are elective (e.g., postsecondary majors and career choices). This situation is even more acute for females from underrepresented racial/ethnic groups and low-income backgrounds. The other main concern is clear evidence of females' less favorable dispositions (e.g., confidence and interest) toward STEM as compared with males. The literature further indicates that females have fewer role models, opportunities, and resources. Females thus have less STEM support and guidance in general than males. Sociocultural factors, such as family and teachers, are identified as having potentially favorable or unfavorable influences on females in STEM, and out-of-school-time (OST) STEM programs are suggested as one strategy for supporting females. Bevan and

Out-of-School-Time STEM Programs for Females, pages 169–174
Copyright © 2017 by Information Age Publishing
All rights of reproduction in any form reserved.

Michalchik (2013) contend, "More STEM OST programs should be supported and made more equitably available" (p. 7).

The chapter authors in this book describe program components considered to be important to the effectiveness of their programs. Those that arise most often are:

- Strategically planned content that strengthens and extends school learning (sometimes involving investigations, real-world projects, or interdisciplinary tasks);
- Carefully chosen instructors that some programs train for their role;
- Active, student-centered instructional approaches (e.g., using hands-on materials, sharing thinking orally and in writing, and engaging in authentic experiences);
- Use of technology to support learning;
- Career awareness efforts;
- Interpersonal connections (e.g., interaction and relationships with peers and role models/mentors in relation to STEM);
- Social and recreational opportunities; and
- Parent/family involvement.

Program staff also strive to create supportive, safe learning environments. Thus, some programs incorporate icebreaker or team-building activities. These efforts fit well with intent to foster favorable dispositions toward STEM, a goal indicated by many programs. Other features noted within some program descriptions in this book are making an effort to reach underserved girls, developing collaborative relationships with local businesses and organizations, showcasing participant learning (e.g., at the end of the experience through participant presentations to families and others), providing resources to teachers, parents, and others for supporting girls in STEM, and conducting follow-up sessions to the main event. Practices described by the authors in this book appear in other literature in the field. These "typical" practices for OST programs include: active recruitment of underrepresented girls; collaborative learning in a supportive atmosphere; use of technological tools to support learning; incorporation of field trips, guest speakers, and participant presentations; staff trainings; parent involvement; and partnerships forged among higher education, local schools, and community businesses and organizations (Davis & Hardin, 2013; McCombs et al., 2012; Mohr-Schroeder et al., 2014; Mosatche, Matloff-Nieves, Kekelis, & Lawner, 2013; Newbill, Drape, Schnittka, Baum, & Evans, 2015; White, 2013).

Conducting programs such as those described in this book are not without challenges. One major concern raised by many of this book's authors is an inability to serve girls longer or serve a greater number of girls by lengthening existing programs, running additional weeks for different participant groups, or conducting the program for a broader age range of participants, as well as reaching girls who have least access to these types of opportunities. Expanding the programs in desired ways tends to be prohibitive because the programs require many resources, such as funding, human investment (e.g., time), and facilities. Similarly, some authors note that procuring funding to conduct these programs is an ongoing challenge. McCombs et al. (2012) say "cost is the main barrier to implementing summer learning programs" (p. 49) and assert that policymakers should support consistent funding for these programs. Paulsen (2013) also raises this key concern. Intra/interpersonal issues are another major area of challenge in running OST programs for girls. These problems mainly involve difficulties some girls have adjusting to the program setting (e.g., experiencing homesickness, learning to interact with a variety of girls, or forming personal connections with peers), behavior issues, and weak self-concepts. Another challenge raised is concern about participant lack of commitment to the program (girls who do not show up or who drop out during the program). One reason participants sometimes miss part of an OST program is because they are involved in other activities at the same time (White, 2013). Effectively supporting all girls is yet another challenge because program participants come from a particularly broad range of abilities and backgrounds.

Data collected on the OST programs in this book are intended to offer insight into program effectiveness, provide information to those who support the programs, and contribute to knowledge in the OST field at large. Many authors call for more research on OST programs to better understand their influence on participant knowledge, skills, and dispositions and to identify program strengths and weaknesses (Constan & Spicer, 2015). Research on OST programs can also play a role in attaining program funding (Constan & Spicer, 2015; Davis & Hardin, 2013; Wilkerson & Haden, 2014). For these reasons, formative and summative evaluation should be incorporated into OST program design (Wilkerson & Haden, 2014).

Data sources across the seven programs featured in this book include surveys containing closed-format and open-ended questions, interviews and focus groups, journal writing and written reflections, work products, observations, and application and attendance data. Data were collected mainly from the girls who participated in the programs but also from parents and program instructors. Data collection was primarily intended to assess potential change in knowledge, skills, and dispositions.

Dispositions, including confidence, attitudes, beliefs, enthusiasm, enjoyment, interest, engagement, perseverance, and perceptions of the usefulness of STEM, were a particularly strong focus in the program evaluations detailed in this book. Indeed, participant dispositions were shown to improve after program participation, and there is some sense among authors that this might be the most important area to target. The second most frequently reported favorable participant outcome is increased knowledge of and interest in STEM degrees and careers. Other participant outcomes reported by two to three programs each are greater STEM knowledge and skills (or at least participant perceptions of such), appreciation for an all-female environment, and enjoyment of the program itself. A number of other outcomes, such as improved leadership and public-speaking skills, were also identified by individual programs. Finally, the data collected for the seven programs in this volume (Volume II is forthcoming) show the following program components to be particularly important: strong academic content; active, student-centered approaches (e.g., hands-on and inquiry-based); and inclusion of social/recreational aspects that allow for engaging the whole self and for more casual participant and staff interactions.

Of chapter authors who suggest future research in the area of OST STEM programs, the long-term influence of these programs is listed most often. The authors recommend longitudinal research centered on investigating program influence on STEM performance, participation, and dispositions at various points in participants' lives, including college attendance and persistence, career choice and career success, and personal lives. This mirrors Mohr-Schroeder et al.'s (2014) statement that "the research literature points to a lack of longitudinal studies following students who attend summer camps into college and subsequently their careers" (p. 292), as well as McCombs et al.'s (2012) call for longitudinal research on OST programs. Another research focus suggested by the book authors is that of gathering more nuanced information about favorable and unfavorable program aspects (e.g., through probing interviews with participants). Finally, some areas of study suggested by individual chapter authors include investigating how the program might influence its instructors and volunteers, how to develop an "ecosystem" of STEM support for girls across the various aspects of their lives (e.g., scholastic, family, community), and how the program might differentially affect different types of girls (e.g., according to race/ethnicity or family income). In relation to the importance of identifying potentially variable program influence on different girls, Ashcraft, Eger, and Friend (2012) note in relation to females and information technology:

> Beyond simply focusing on gender, consider the importance of intersectional research and programs that explore multiple intersections of youth's identities for computing pedagogy (e.g., race, class, gender, and sexuality). Diversity of voices and experiences will help not only in the production of richer research but also a richer U.S. computing workforce. (p. 45)

Other research suggested for OST programs in the professional literature includes the effectiveness and proper use of various types of technological tools (Paulsen, 2013) and how and to what degree learning takes place within and across different settings in and out of school (Afterschool Alliance, 2013; Bevan & Michalchik, 2013). McCombs et al. (2012) recommend studies using randomized trials to compare program participants and nonparticipants.

Given the potential role out-of-school-time programs might play in bolstering student learning, especially for underrepresented groups, we believe OST programs should be the subject of greater study and discussion. Further, given that existing data on OST programs demonstrate favorable outcomes, we call for greater funding to increase out-of-school-time program offerings. Special efforts should be made to engage underrepresented and underserved youth, who might need these programs the most. Such is the focus of this book, which describes ways OST programs can support females in science, technology, engineering, and mathematics.

REFERENCES

Afterschool Alliance. (2013). *Defining youth outcomes for STEM learning in afterschool.* Retrieved from http://www.afterschoolalliance.org/STEM_Outcomes_2013.pdf

Ashcraft, C., Eger, E., & Friend, M. (2012). *Girls in IT: The facts.* Boulder, CO: National Center for Women & Information Technology, University of Colorado. Retrieved from https://www.ncwit.org/resources/girls-it-facts

Bevan, B., & Michalchik, V. (2013, Spring). Where it gets interesting: Competing models of STEM learning after school. *Afterschool Matters,* 1–8.

Constan, Z., & Spicer, J. J. (2015). Maximizing future potential in physics and STEM: Evaluating a summer program through a partnership between science outreach and education research. *Journal of Higher Education Outreach and Engagement, 19*(2), 117–136.

Davis, K. B., & Hardin, S. E. (2013). Making STEM fun: How to organize a STEM camp. *TEACHING Exceptional Children, 45*(4), 60–67.

McCombs, J. S., Augustine, C., Schwartz, H., Bodilly, S., McInnis, B., Lichter, D., & Cross, A. B. (2012). Making summer count: How summer programs can boost children's learning. *The Education Digest, 77*(6), 47–52.

Mohr-Schroeder, M. J., Jackson, C., Miller, M., Walcott, B., Little, D. L., Speler, L., ... Schroeder, D. C. (2014). Developing middle school students' interests in STEM via summer learning experiences: See Blue STEM Camp. *School Science and Mathematics, 114*(6), 291–301.

Mosatche, H. S., Matloff-Nieves, S., Kekelis, L., & Lawner, E. K. (2013, Spring). Effective STEM programs for adolescent girls: Three approaches and many lessons learned. *Afterschool Matters,* 17–25.

Newbill, P. L., Drape, T. A., Schnittka, C., Baum, L., & Evans, M. A. (2015, Fall). Learning across space instead of over time: Redesigning a school-based STEM curriculum for OST. *Afterschool Matters,* 4–12.

Paulsen, C. A. (2013, Spring). Implementing out-of-school time STEM resources: Best practices from public television. *Afterschool Matters,* 27–35.

White, D. W. (2013). Urban STEM Education: A unique summer experience. *Technology and Engineering Teacher, 72*(5), 8–13.

Wilkerson, S. B., & Haden, C. M. (2014, Spring). Effective practices for evaluating STEM out-of-school time programs. *Afterschool Matters, 19,* 10–19.

SELECTED OST RESOURCES

JOURNAL ARTICLES

An increasing number of resources provide information on OST programs. Selected research articles may be found in the end references of individual chapters of this book. Several helpful articles that provide practical information on conducting OST programs include:

- Augustine, C. H., & McCombs, J. S. (2015). Summer learning programs yield key lessons for districts and policymakers. *The State Education Standard, 15*(1), 11–19.
- Davis, K. B., & Hardin, S. E. (2013). Making STEM fun: How to organize a STEM camp. *TEACHING Exceptional Children, 45*(4), 60–67.
- Mosatche, H. S., Matloff-Nieves, S., Kekelis, L., & Lawner, E. K. (2013, Spring). Effective STEM programs for adolescent girls: Three approaches and many lessons learned. *Afterschool Matters,* 17–25.
- Wiest, L. (2008). Conducting a mathematics camp for girls and other mathematics enthusiasts. *The Australian Mathematics Teacher, 64*(4), 17–24.
- Wilkerson, S. B., & Haden, C. M. (2014, Spring). Effective practices for evaluating STEM out-of-school time programs. *Afterschool Matters, 19,* 10–19.

Out-of-School-Time STEM Programs for Females, pages 175–176
Copyright © 2017 by Information Age Publishing

PROGRAM DIRECTORIES

A number of databases catalog out-of-school-time programs across the United States. Two key directories are:

- Out-of-School Time Program Research & Evaluation Database & Bibliography (Harvard Family Research Project, Harvard Graduate School of Education): http://www.hfrp.org/out-of-school-time/ost-database-bibliography
- STEM Program Directory (National Girls Collaborative Project): http://www.pugetsoundcenter.org/ngcp/directory/index.cfm

ABOUT THE CONTRIBUTORS

Lindsay Augustyn, MA, is the outreach and communications director for the Center for Science, Mathematics and Computer Education at the University of Nebraska–Lincoln. She focuses primarily on editing, design, recruitment, and conference development. She has been involved with All Girls/All Math since 2010.

Heather Glynn Crawford-Ferre, MEd, is a doctoral candidate at the University of Nevada, Reno. She serves as a consultant for the Nevada Department of Education and has worked with the Northern Nevada Girls Math & Technology Program since 2007. Her scholarly interests are mathematics education, educational equity, teacher education, and out-of-school time programming.

Diana B. Erchick, PhD, is a professor at the Ohio State University at Newark. Her scholarly interests are equity and social justice in mathematics education, and mathematics teacher education. She has directed the Matherscize Camp for Middle Grades Girls since she developed it in 1999.

Gwendolen ("Wendy") Hines, PhD, is a professor emeritus at the University of Nebraska–Lincoln (UNL). Her research area is dynamical systems. She and Judy Walker co-developed the All Girls/All Math camp in 1997 and co-directed it for five years. After that, she was the sole director until her retirement in 2014.

Out-of-School-Time STEM Programs for Females, pages 177–179
Copyright © 2017 by Information Age Publishing
177

Angie Hodge, PhD, is an associate professor at the University of Nebraska–Omaha. Her scholarly interests are mathematics education, gender equity in STEM fields, inquiry-based learning, and professional development for mathematics teachers. She helped found the EUREKA!-STEM program in 2012 and has been active in leadership in the program since then.

Jeanne Houck, PhD, is a historian who specializes in developing cultural programs with educational components for K–12 audiences. As director of foundation relations at the Intrepid Sea, Air & Space Museum, she builds strategic partnerships that support educational programs.

Linda Kekelis, PhD, consults on programs and research related to STEM, particularly on increasing access for girls and underrepresented youth. She collaborates with girl-serving organizations, participates on advisory boards, and works with partners to build their capacity for quality STEM programming. She is the founder and former CEO of Techbridge.

Lynda Kennedy, PhD, is Vice President of Education & Evaluation for the Intrepid Sea, Air & Space Museum. Working between school and cultural settings for two decades, her research interests center on the role of cultural institutions, such as museums, zoos, science, and arts centers in the educational ecosystem.

Sheri Levinsky-Raskin, MAT, is the Assistant Vice President of Education & Evaluation at the Intrepid Sea, Air & Space Museum, where she has served since 2005. She has been an active contributor to the museum education field since 1996 and involved with GOALS for Girls since its inception in 2007.

Michael Matthews, PhD, is an associate professor at the University of Nebraska–Omaha. His scholarly interests are mathematics education, inquiry based learning, educational technology, and teacher education. He has been involved with the Eureka!-STEM program since its inception in 2012.

Kerry F. McLaughlin, MA, earned a bachelor's degree in history from York College of Pennsylvania and a master's degree in museum studies from Seton Hall University. She has worked with out-of-school-time audiences at the Intrepid Sea, Air & Space Museum for 12 years.

Shihadah ("Shay") Saleem, MS, is a senior educator at the Intrepid Sea, Air & Space Museum. Her expertise and interests lie in STEM program development for girls, teachers, and communities. She has been the coordinator and director of GOALS for Girls since co-founding the program in 2007.

Jafeth E. Sanchez, PhD, is an assistant professor at the University of Nevada, Reno. Her scholarly interests are focused on educational leadership, as well as gender and ethnic equity, educational outreach, and P16 alignment. She worked with the Northern Nevada Girls Math & Technology Program from 2006 to 2012.

Hortensia Soto-Johnson, PhD, is a professor in the School of Mathematical Sciences at the University of Northern Colorado. Her research focuses on the teaching and learning of undergraduate mathematics via an embodied cognition lens. She enjoys giving back to the community through programs such as *Las Chicas de Matemáticas.*

Amelia Squires, MS, graduated with a master's degree in Elementary Education with a concentration in STEM Education from the University of Nebraska Omaha (UNO) in 2015. Since 2014, she has worked as the STEM Outreach Coordinator at UNO. She is the Project Director of the UNO Eureka!-STEM summer program.

Lynda R. Wiest, PhD, is a professor at the University of Nevada, Reno. Her scholarly interests are mathematics education, educational equity, and teacher education. She has directed the Northern Nevada Girls Math & Technology Program since she developed it in 1998.

Made in the USA
Columbia, SC
02 September 2017